T0192818

Internet of Things and M2M Communication Technologies

Veena S. Chakravarthi

Internet of Things and M2M Communication Technologies

Architecture and Practical Design Approach to IoT in Industry 4.0

 Springer

Veena S. Chakravarthi (iD)
Sensesemi Technologies Pvt. Ltd.
Bangalore, Karnataka, India

ISBN 978-3-030-79274-9   ISBN 978-3-030-79272-5   (eBook)
https://doi.org/10.1007/978-3-030-79272-5

This Springer imprint is published by the registered company Springer Nature Switzerland AG
The registered company address is: Gewerbestrasse 11, 6330 Cham, Switzerland

Dedicated to product designers.

Foreword[1]

The advent of Industry 4.0 ushers in a new era where modern technology and connected devices are used in business and society. Manufacturing practices are becoming automatic at breakneck speed. Cities, buildings, and homes are becoming smart. Vehicles are becoming intelligent and autonomous. Thanks to Internet of Things (IoT); smart services are sprouting everywhere. The growth of IoT is further fueled by cutting-edge technologies like 5G, AI, and Big Data.

Research agencies predict that the IoT market will embrace 50 billion connected devices by 2025. The making of smart things has never been more complex and it is only going to get more sophisticated. Every device needs to be smaller, smarter, more powerful, and more energy-efficient than the existing ones. Simultaneously, we take cognizance that Generation Zs are hyper-cognitive and true digital natives. Creating self-learning connected machines by instilling intelligence to IoT excites them. They are the consumers of smart gadgets and the first generation of technopreneurs.

This is an excellent time to be working in the automation industry that is driven by IoT and machine-to-machine (M2M) connectivity. Any aspiring IoT engineer, whether working in a startup or in a large corporate, needs to become well versed in

[1] **Dr. Aloknath De** is the senior Vice President and chief technology officer of Samsung R&D Institute India, Bangalore, since 2011 when he initiated 5G research activities as his first assignment. As corporate vice president of Samsung Electronics, South Korea, he has also been responsible for Samsung's IoT Data Platform and Intelligent Services. Dr. De received his B.Tech. from Indian Institute of Technology (IIT), Kharagpur, M.E. from Indian Institute of Science (IISc), Bangalore, and Ph.D. from McGill University, Montreal. Prior to Samsung, he has held progressive responsibilities in Bharat Electronics, Nortel Network (Montreal), Hughes, and ST-Ericsson. Additionally, he served as an AICTE-INAE visiting professor in IIT-Roorkee and an adjunct professor in IIT-Delhi. He has 25+ patents granted/filed and 50+ research publications. For technology leadership, Dr. De has received many accolades including Alexander Graham Bell Prize (Canada), IETE Memorial Award, IDC Insights Award, Nasscom AI Game Changer Award to name a few. His meticulous drive has made Samsung India center an Innovation Hub with repeat recognition through "National IP Award" and "Great Place to Innovate" award.

the art of hardware, software, protocols, and product design. Dr. Veena Chakravarthi has taken an *ab initio* perspective while authoring this book. In my parlance, IoT stands for "Integration of Technologies." For an IoT system, stakeholders are many. It is not easy to write a book that intertwines multiple technologies toward a meaningful value-adding service.

This book deals with the complexity involved in developing multi-domain IoT solutions. It covers necessary fundamentals in terms of theories, relevant standards, as well as design methods of IoT. It also delineates short-range wireless technologies and M2M communication protocols. This is the second book from Dr. Veena, the first one being in the area of system-on-chip (SoC). Dr. Veena has seen both the worlds of academia and industry. She understands well the need of uninitiated in a topic and knows equally the art of initiating them with step-by-step practical methods.

I am excited about the future of our industry and how far IoT can spread its wings. There is no doubt that the IoT industry is creating new markets, paving ways for innovation in making everything smart and connected. AI-infused IoT (AIoT) is at the heart of the advancements in healthcare, transportation, telecommunication, manufacturing, logistics, and other verticals. The book provides reference design materials and illustrative examples pertaining to analyzing health and monitoring environment.

This book is a thoughtful guide for any aspiring IoT engineer and product designer. People with experimental and do-it-yourself mindset will immensely benefit. I am sure that the book will aid significantly the next-generation engineers, technologists, and innovators.

Senior Vice President and CTO-Samsung   Aloknath De
Bangalore, India

Foreword[1]

I have had the pleasure of working with Dr. Veena Chakravarthi multiple times and in different contexts during my professional life of more than three decades. We started our careers in the same company and on the same day! Our paths have crossed frequently, and she has spent almost a decade of her rich experience working with me in building a successful new engineering organization within a larger multinational company to develop a (then) state-of-the-art access system. We have also been part of multiple industry forums where I have had the privilege of tapping on her theoretical strength and practical insights.

Internet of Things (IoT) and machine-to-machine are revolutionary concepts that are integral to the imminent Industrial Revolution 4.0. This makes it important for all industry planners as well as electronics and computer science engineers to be conversant with these topics to stay relevant in the new times. However, the requirements of IOT and M2M are diverse and

---

[1] **M.N. Kumar**, Chief strategy officer (CSO) and head of engineering, comes with more than three decades of experience in multiple successful product startups and MNCs. He has nurtured and built teams to successfully design, deliver, and support SoCs and SoC-based systems to Tier-1 customers worldwide.

He has worked at Microelectronics Group at ITI, Bangalore, and ST Microelectronics in Agrate, Italy, before joining Arcus Technology, possibly India's first fabless Chip Company, which was acquired by Cypress Semiconductors. In 2000, he left Cypress Semiconductor to co-found vEngines, an early innovator of packet voice processors for use in VoIP, etc. vEngines was acquired by Centillium Communications, where he lead engineering for various product lines like Voice Processors, xDSL, EPON, Home Gateway, etc. After Centillium merged with Transwitch, he was the head of the Bangalore Technology Center. In 2012, he started and led the India Operations of Aquantia Inc. Aquantia had its IPO in 2017 and was acquired by Marvell Semiconductors in 2019, where he was Sr. Director, before joining LeadSoC.

He holds a master's degree in microelectronics from IIT-Bombay. He is a Sr. member of IEEE and past chairperson of the Nano Technology Council—IEEE Bangalore.

span multidisciplines and, therefore, difficult to collect and assimilate easily. This book is a much-needed text for those who want to study the various relevant components of these technologies and design trade-offs in a single concise book. Such a book must provide enough practical information to enable an early takeoff in this field for a new entrant. But it must also provide theoretical information to ignite interest in the reader to pursue further exploration on the areas. This is a fine balance that requires an author who can do justice to both, seemingly opposing, requirements. Moreover, such a text must provide real case studies for the readers to appreciate the approach and possibilities.

Dr. Veena Chakravarthi has a vast spectrum of experience spanning industry and academia. She has experience in multiple domains of applications for which she has designed solutions, etc. which provide the ideal mix for a book of this type. She has the advantage of having successfully deployed new green-field products, including IoT products for health. Her experience in the deployment of new products, including in her start-ups and post-deployment support for these products give an extra edge as this experience has no substitute as this gives deep knowledge of field and deployment issues, which cannot otherwise be accessed. This knowledge in turn helps to iteratively tune future products for even higher efficacy. Her publications in communications, semiconductor design and IoT give ample evidence of her theoretical moorings and her real-life product experience enables her to be an apt author for a book like this. Her book on system-on-chip (SOC) design has been a phenomenal success and is a prescribed text in many universities apart from being popular with practicing VLSI engineers.

I am happy that this book has become available at this stage of takeoff of IoT and M2M solutions. This gives an easy fillip to those desiring to get IoT and M2M Ready for the industry. This book is also a wonderful book for practicing engineers due to its design cases and template-type approach which will help most engineers map many of their problems within the confines of the matter discussed.

The methodical buildup of the material with an easy-to-understand approach makes this text accessible both as a strong concrete theoretical text to know underlying principles and as a practical approach to accessing selective information at any stage of design and development. These are the same reasons for which her book on SOC Design is a super-successful book with both students and practicing engineers.

In conclusion, I strongly welcome this much-needed book, dealing with multiple disciplines, which is both a useful primer for newcomers into the IoT and M2M domains and a strong reference text for existing designers. Dr. Veena Chakravarthi has undertaken a much-desired task that can help introduce these new areas easily to hitherto hesitant engineers and also act as a good supplementary support material for the existing IoT and M2M design community.

Chief Strategy Officer (CSO) & Head   M. N. Kumar
Engineering
LeadSOC
Bangalore, India

Preface

Industry 4.0 is the fourth generation of the evolution that the industry is undergoing. It is driven by the Internet of Things and connectivity. Internet of Things in particular is promising a major shift in lifestyles by connecting the things around us onto the Internet and bringing a heightened sense of comfort. This creates new principles of design for us to organize ourselves around smart things and new lifestyles.

Industry 4.0 is driving a radical shift in automation of manufacturing and processes, promising huge productivity and efficiency.

Having worked in semiconductor design and wireless industry for over two decades, it was my strong desire to pass on the knowledge of Internet of Things (IoT) and M2M communication technology to the next generation. And so I conceived the idea of writing a book on *Internet of Things and M2M Communication Technologies: A Practical Approach to Architecture and Design.*

The book intends to present a comprehensive overview of the design methodology, development environment, protocols, and requisite skills that are required for the design and development of IoT-based product design.

It ensures that engineers are aware and are able to contribute effectively in the much talked about Industry 4.0 revolution.

While the book is targeted to all engineers (Electrical, Electronics and Communication, Computer Science, and Product Design) who aspire to be IoT designers, it is also a valuable reference guide for professional designers who are part of development teams in IoT, manufacturing, and fabrication houses.

The book aims to give readers a comprehensive idea of what one has to do as a hardware, software, and product designer. It emphasizes on the protocols and standard regulations to be followed, and building necessary security features in the product design. IoT product design is multidisciplinary, and the book covers necessary skills from different engineering portfolios and the challenges that one can anticipate during product design. This information is based on my experiences in the product industry and academics for the past 30 years.

Typically, most engineering students aspire to build products as solutions during and after their graduation. Unfortunately for them, they usually do not possess the requisite skills and know design techniques to circumnavigate the challenges they will face in realizing the products.

Meanwhile, young product designers struggle to see the big picture of the IoT product design process. It is not practical for one person to work on all areas of the product design and development process. This book is my attempt to provide answers to both groups, so that they can plan, understand, and equip themselves with necessary skill sets in product design discipline of their interest. The design case relevance in every chapter helps the reader realistically visualize problems and solutions encountered during IoT system design.

The target audience for this book is engineering students who are pursuing a degree in electrical, electronics and communication, computer science, and allied branches like biomedical, biotechnology, product engineering, AI, ML and Data Science etc. Also, engineers in early stages of their career working in product companies can refer to the book for a complete understanding of the product design process.

Though the book covers the complete spectrum of the topics relevant to IoT design, it is good to have a fundamental understanding of microcontroller and communication technologies as it is a prerequisite to follow the contents of the book.

Though India has been encouraging the product and manufacturing industry with various schemes of "Startup India," "Make in India," it is facing an acute shortage of product engineers as a large number of fresh engineering graduates from universities are not readily deployable for the manufacturing and product industry.

Statistics show that 13% of the jobs are held by the product industry with close to 300,000 jobs available in product and manufacturing in all sectors. There is a demand for engineers who can design and develop products in the coming years.

In this context, a product design engineer with multidisciplinary skillset has a promising and bright future ahead and can expect a challenging and rewarding career. Globally, Industry 4.0 is the reality, and every engineer has a role to play in it.

The design productivity gap—a shortage of skilled manpower to make everything around us smart and intelligent—is real. Hence, there is a need to develop skill sets to suit the product design and development jobs to bridge this gap. This book is an endeavor in this regard.

It would not have been possible to realize this project without the support of many of my friends, colleagues, and family. First, I wish to thank my father **R S Chakravarthi**, a noted journalist and Rajyotsava awardee from Karnataka, India, whose literary gene was responsible for harboring my desire to write books. My heartfelt thanks to my loving family, my husband **Dr. K S Sridhar**, Registrar, PES University, my dear sons—**K S Abhinandan** and **K S Anirudh**—and my loving daughter-in-law **Shradha Narayanan**.

I am indebted to my colleague, **Dr. M S Suresh**, scientist from ISRO, who patiently read each of my chapters and offered line-by-line reviews.

I wish to thank my ex-colleague **Rashmi Venkataramaiah** for the reference design. My steadfast team comprises **Dr. Bindu** and **Dr. Yasha Jyothi M Shirur** for reviewing chapters. Thanks to them.

I am also grateful to my company **Sensesemi Technologies** and before that the semiconductor industry for having embraced me so warmly. And I am mighty thankful to **Dr. Aloknath** De, CTO, Samsung India, and **Mr. M N Kumar**, CSO, LeadSoc, India for taking time out of their busy schedules to write the foreword for this book.

Last but not the least, I thank my superpower, who gives me the motivation and constant energy to take up projects beyond my capability and make them happen.

About the book organization, Chapters 1 to 9 cover in detail the Internet of Things (IoT), including Industrial Internet of Things (IIoT) related topics. It deals with the complexity involved in developing multi-domain IoT product covering necessary fundamentals in terms of theories, relevant standards, and design methods. A simple product design for environmental monitoring is presented as a reference design. Chapters 10 to 12 of the book cover wireless technologies and M2M communication protocols and Chapter 13 deals with database management and analytics. Chapter 14 presents the design for manufacturability and different business models for those who want to take a step further to understand the business around these technologies. Chapter 15 has a question bank corresponding to each chapter to assess their understanding in these topics and try some design examples. The design approach to the healthcare ecosystem is presented in Chapter 15, for those of you who want to solve real-life complex ecosystem problems. The references provided in the form of website links or cited in the description in each chapter are for more detailed reading, in addition to whatever is covered.

I will be very happy if users find each chapter useful and try out the reference design. I am curious about your feedback and criticisms. I am sure this will go a long way in bettering this book.

Thank You.

Bangalore, India   Veena S. Chakravarthi

Contents

Abbreviations

3G	Generally referred to third-generation wireless mobile telecommunication technology
6LowPAN	IPv6 over low-power wireless personal area networks
ADC	Analog-to-digital converter
AES	Advanced Encryption Standard
AI	Artificial intelligence
API	Application programming interface
ARP	Address Resolution Protocol
AWS	Amazon web services
BIOS	Basic input–output system
BSP	Board support package
BT-LE	Bluetooth Low Energy (BLE)
CCD	Charge-coupled device
CDK	Cloud development kit
CDSCO	Central Drugs Standard Control Organization
CMOS	Complementary metal oxide semiconductor
CoAP	Constrained Application Protocol
CoRE	Constrained RESTful environment
CSA	Cyber security act
CSE	Common service entity
DAC	Digital-to-analog converter
DBMS	Database management system
DDB	Distributed database
DDMS	Dynamic database management system
DFM	Design for manufacturability
DHCP	Dynamic host configuration protocol
DLT	Distributed ledger technologies
DNS	Domain name system
DSP	Digital signal processing
DTLS	Datagram transport layer security
ECG	Electrocardiogram

EEPC	Engineering Export Promotional Council of India
EIS	Electronic image stabilization
EoF	End of frame
ERP	Enterprise resource planning
ETSI	European Telecommunications Standards Institute
EU	European Union—Medicines and Healthcare products Regulatory Agency (MHRA)
FCC	Federal Communications Commission
FDA	Food and Drug Administration
FOV	Field of view
FR	Frame rate
FSSAI	Food Safety and Standards Authority of India
GNSS	Global navigation satellite system
GPIO	General-purpose input–output
GPS	Global positioning system
GPU	General processor Unit
GUI	Graphical user interface
HAL	Hardware abstraction layer
HDR	High definition range
HMD	Head-mounted devices
HMI	Human-machine interface
HTML	Hypertext markup language
HTTP	Hypertext transfer protocol
I 4.0	Fourth-generation Industry Revolution
I2C	Inter-integrated circuits communication; In IoT context, it is a method for communicating between devices such as sensors, displays, and other peripherals and a microcontroller.
IDE	Integrated development environment
IDMS	Industrial data management system
IEC	International Electrotechnical Commission
IEEE	Institute of Electrical and Electronics Engineers, off late referred as just IEEE
IETF	Internet Engineering Task Force
IIoT	Industrial Internet of Things
IoT	Internet of Things
IoTA	Integration of Two Access, a distributed ledger designed to record and execute transactions between machines and devices in the Internet of Things (IoT) ecosystem
IP	Internet Protocol
IR	Infrared radio frequency
ITU-T	International Telecommunication Union
JSON	Javascript object notation
JTAG	Joint Test Action Group
LAN	Local area network
LCD	Liquid crystal display

LLN	Low-power lossy networks
LoRAWAN	Long-range wireless area network
LR	Low rate
LR-WPAN	Low-rate wireless personal area networks
LTE	Long-term evolution
LWM2M	Lightweight machine-to-machine
M2M	Machine to machine
MAC	Media access controller
MCU	Microcontroller unit
MEMS	Microelectromechanical systems
MHRA	Medicines and Healthcare products Regulatory Agency
MIPS	Million instructions per second
MiTM	Man-in-the-middle attack
ML	Machine learning
MQTT	Message Queuing Telemetry Transport
MTU	Maximum transfer unit
NAN	Neighbor area network
NB-IoT	NarrowBand-Internet of Things
NFC	Near-field communication
NISD	European Union's Network and Information Security Directive
NIST	National Institute of Science and Technology
NRE	Nonrecurring engineering
NVM	Non-volatile memory
OCF	Open Connectivity Foundation
OEM	Original equipment manufacturer
OIC	Open Interconnect Consortium
OIS	Optical image stabilization
OOP	Object-oriented programming
OPC	Open platform communication
OS	Operating system
OSI	Open systems interconnection
OTA	Over the air
PAAS	Platform as service
PDAF	Phase detection auto focus
PDK	Process design kit
PLC	Programmable logic controller
PoC	Proof of concept
PUF	Physical unclonable functions
PWM	Pulse width modulation
QFN	Quad flat no-lead
QoS	Quality of service
RAM	Random access memory
RDMS	Relational database management systems
ReDoS	Regular expression denial of service
REST	Representational state transfer

RF	Radio frequency
RFID	Radio frequency identification
RoT	Root of trust
RPL	Routing protocol for low-power and lossy networks
RTOS	Real-time operating system
RTT	Round trip time
RTU	Remote terminal unit
SBC	Single-board computer
SCADA	Supervisory control and data acquisition
SCCB	Serial camera control bus
SDE	Software development environment
SDH	Synchronous digital hierarchy
SDIO	Secure Digital Input Output
SDK	Software development kit
SG	Study group
SIL	Safety integrity level
SMS	Short message service
SoAP	Simple Object Access Protocol
SOC	System on chip
SONET	Synchronous optical network
SPI	Serial Peripheral Interface
SQL	Structured query language
SSD	Solid state drives
STP	Spanning Tree Protocol
T/TCP	Transactional Transmission Control Protocol (T/TCP)
TCP	Transmission Control Protocol
TCP/IP	Transmission Control Protocol/Internet Protocol
TFTP	Trivial File Transfer Protocol
TLS	Transport layer security
ToS	Type of service
TRAI	Telecom Regulatory Authority of India
TTL	Time to live
UART	Universal asynchronous receiver-transmitter
UDP	User datagram Protocol
UDP	User datagram protocol
UE	User equipment
UI and ML	User interface and machine learning
UI	User interface
UPM	Useful Packages and Modules
URI	User resource identifier
URL	Uniform resource locator
URN	Uniform resource name
USB	Universal serial bus
USP	Unique selling proposition
VLAN	Virtual LAN

VLSI	Very large scale integration
VMX	Virtual machine extensions instructions
VoIP	Voice over Internet Protocol
VPN	Virtual private network
WiFi	Wireless Fidelity; also referred to wireless local area network technology
WLAN	Wireless local area network
WSN	Wireless sensor network
XML	eXtensible Markup language
XMPP	Extensible Messaging and Presence Protocol

About the Author

Veena S. Chakravarthi is a Bangalore-based technologist, system-on-chip architect, and educator. Over a career spanning three decades, she has spawned several VLSI design and incubation centers and managed several high-performance tech-teams at ITI Limited and across various MNCs like Mindtree Consulting Pvt. Ltd., Centillium India Pvt. Ltd., Transwitch India Pvt. Ltd., Ikanos Communications Pvt. Ltd., Pereira Ventures, Asarva Chips and Technologies, Saankhya Labs, and Prodigy Technovations Pvt. Ltd. She has been the research head and an adjunct professor in the Department of Electronics and Communication Engineering, at BNM Institute of Technology, Bangalore. She also holds a PhD from Bangalore University and an MPT certification from IIM Bangalore.

Chapter 1
Internet of Things: An Introduction

1.1 Industry 4.0

The objective of science and technology is to improve the standard of living, by improving economies of societies. In that regard, ongoing fourth generation of the Industrial revolution draws relevance to maximize the benefits of technologies and avoid potential risks. This ongoing revolution was named Industry 4.0 or I4.0 by the World Economic Forum in the year 2016. Internet technology is responsible for this growth and is the key technology enabler which has propelled innovation and economies over the past few decades. It still continues to be the primary enabling technology for the current and the future advancements.

1.2 A Brief History of Industrial Revolutions

Invention of steam engine triggered the First Industrial Revolution in the eighteenth century. It gave birth to mechanization of production which was earlier done manually. What was produced manually by spinning wheels were automatically produced using steam engines, and the productivity was found to be eight times higher. Development of steam locomotives further triggered movement of people and goods to longer distances increasing the reach.

The *Second Industrial Revolution* was due to the invention of electricity. This began in the nineteenth century through the use of electricity and assembly line production. The part processing at each stage was adopted in sequence till the entire process was completed. This reduced cost of production drastically and made it faster.

The *Third Industrial Revolution* was due to memory-configured automation and computer controls without human intervention. This automated the entire production. Few examples of this are robots, Computer numerical control (CNC) machines, etc. with no or minimum human intervention.

The ongoing *Fourth Industrial Revolution (I4.0)* is due to amalgamation of information and communication technologies, which is being built over the Third Industrial Revolution. The production systems with computers are inter-networked to communicate and collaborate to achieve complete automation, the next step in production automation. The networking of all systems leads to "cyber-physical production systems," and therefore, smart factories, where production systems, components, and people communicate via a network to make autonomous production. I4.0 is predicted to have the capability to foresee possible future risky scenarios and act upon them with artificial intelligence (AI) to mitigate them. Figure 1.1 shows the history of Industrial Revolutions. Industry 4.0 creates rapid and disruptive changes in every line and sector.

1.3   Technology Enablers

In *I4.0*, the machines manage and control themselves in a production process. The machines are computer controlled, communicating with each other to make intelligent decisions. This advanced technology is called *Internet of Things*. Other enabling technologies responsible for I4.0 revolution apart from the Internet of Things are *artificial intelligence* and *machine learning, sensor technologies, MEMs* and *nanotechnologies*, and *machine-to-machine (M2M) communication* technologies.

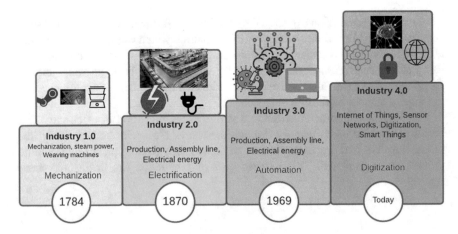

Fig. 1.1 History of Industrial Revolutions

1.4   Internet of Things

All of us interact with the physical world to make it comfortable so that people are more productive at work, happier, and smarter at all times. The *Internet of Things (IoT) is one of the many means which is technology driven to achieve this.*
Examples are:

1. If you feel hot in summer, there should be a means to automatically sense it, control the temperature of the surroundings, and make it cooler. This, in conventional systems, makes the fans or air conditioning systems smarter to sense the surroundings and control them such that the surrounding environment is made cooler and comfortable.
2. When you decide to go shopping, a car awaiting outside opens the door for you, controls the inside of the car to suit your mood, plays a soothing song of your liking, and drives to drop you near the shopping mall.
3. When you step into a dark room, the device senses your entry and turns on the light and, as you leave the room, automatically switches off the lights.
4. The water level controller of overhead tank senses the water level had gone lower in the tank and turns on the motor to fill the water on the overhead tank.

Few of the Internet of Thing (IoT) device applications are shown in Fig. 1.2.

Fig. 1.2 IoT applications

The Internet of Things (IoT) is a physical device, which communicates with other physical devices on the Internet. This technology enables communication of the information across devices for better control or derives greater meaning from the individual device information. In short, IoT connects any physical thing, people, animal, processes, and plants, through the Internet, without the need for people to constantly maintain the connection or transfer of data manually. It is in a way automatic information transfer. The Internet of Things (IoT) is a key enabler for many emerging and future "smart" applications and technology in various technology markets. This ranges from the connected consumer to smart home and buildings, e-health, smart grids, next-generation manufacturing, and smart cities. It is therefore predicted to be one of the significant drivers of growth.

1.5 IoT Concept

IoT is a concept viewed differently depending on perspectives but depending on integration of multiple technologies can become the smart solution to a challenge in any domain. The IoT concept is shown in Fig. 1.3.

IoT Analytics Research 2018 predicts that by 2025, the number of things connected will exceed 20 billion and the data generated by these connected things will exceed the data generated by connected people as shown in Fig. 1.4.

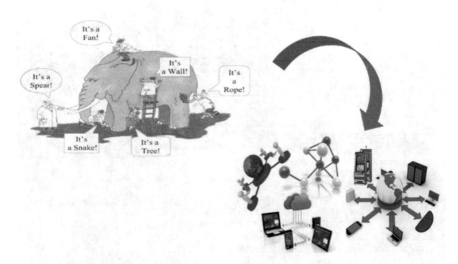

Fig. 1.3 IoT concept (Courtesy: IEEE-SA ETSI M2M Workshop Sophia Antipolis, France 10, December 2014)

Number of global active IoT Connections (installed base) in Bn

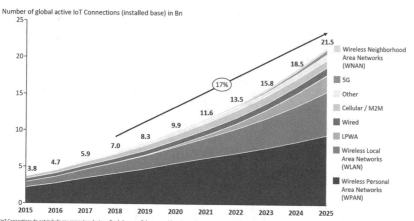

Note: IoT Connections do not include any computers, laptops, fixed phones, cellphones or tablets. Counted are active nodes/devices or gateways that concentrate the end-sensors, not every sensor/actuator. Simple one-directional communications technology not considered (e.g., RFID, NFC). Wired includes Ethernet and Fieldbuses (e.g., connected industrial PLCs or I/O modules); Cellular includes 2G, 3G, 4G; LPWAN includes unlicensed and licensed low-power networks; WPAN includes Bluetooth, Zigbee, Z-Wave or similar; WLAN includes Wi-fi and related protocols; WNAN includes non-short range mesh; Other includes satellite and unclassified proprietary networks with any range.
Source: IoT Analytics Research 2018

Fig. 1.4 IoT connectivity trend (Source: IoT Analytics Research 2018)

1.6 Characteristics of Internet of Things

The main characteristics of IoT devices are to capture or collect the data needed, capable of process the collected data them, and derive meaningful information from the collected and processed data, communicate to the other devices on the network to collaborate or process them further, extract control or output parameters, and control the devices or display the processed data as the case may be.

The Internet of Things are characterized by the following features:

- IoT device can sense/capture and monitor physical parameters.
- IoT device can sense/capture and monitor body vital parameters.
- IoT device can sense/capture and monitor activities of the subject (humans or animals).
- IoT device is an easy to use, install, or deploy on body or on field depending on applications.
- IoT devices can process the raw data captured/monitored to derive meaningful information.
- IoT device can aggregate related parameters to derive bigger sense out of the information aggregated.
- IoT device can identify, analyze, and process using historical data of similar nature.
- They can predict trend of the aggregated data analyzing the current and historical data set.

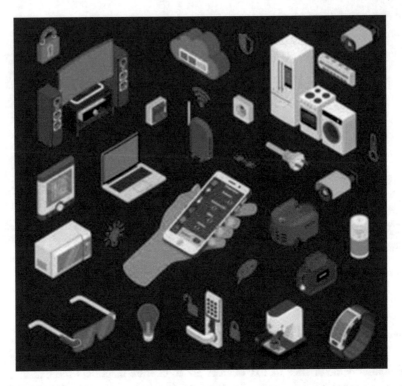

Fig. 1.5 IoT devices

- IoT device can alert or communicate to the stakeholder about the data set quality and sometimes suggest corrective action depending on the knowledge base they have gathered earlier on similar data set.
- IoT device can automatically control the surroundings or related parameters to make the data set normal.
- All of the above activities in IoT device can happen with or without human intervention.

A few examples of IoT devices are shown in Fig. 1.5.

Some examples of IoT devices include smart watch, Fitbit, Google Glass, smart energy meters, smart home appliance controllers, smart home automation systems, smart city devices like traffic controllers, smartphones with health trackers, health watches, smart street light controllers, smart software laptop/mobile applications, smart footwears, smart wearables, and environment trackers. Table 1.1 shows the feature map of some of the commercially available IoTs with their characteristic features listed in this section.

Table 1.1 Feature map of commercially available IoTs with characteristic features

Sl. no.	IoT device	Commercial features
1	NoiseFit smart watch	• Activity tracking • Smart notifications • Heart rate tracking • Camera control • Music player control • Sleep tracking • Steps and calorie tracking
2	Mi Smart Band	• Music player control • Swim style tracking • Health and wellness tracking • 24/7 heart rate monitoring • Style the band display based on mood • Message and call notifications • Social media notifications
3	Apple smart watch	• Automatically synchronizes with Apple iPhone • Selected applications can be shared with watch • It can independently monitor activities
4	Krome automated home lighting system	• Automates all home appliances and systems • Home lighting system • Home theatre everywhere • Ambient surroundings • Monitor energy consumption and control
5	IOTA	• Home energy monitoring and control
6	Cobalt digital door lock	• Automatic misalignment correction • Sideload release • High security • Dual monitoring
7	Vuzix Blade Upgraded Smart Glasses	• Self-contained augmented reality smart glasses • Deliver a hands-free connection of the digital world to the real world • Providing unprecedented access to location-aware information • Data collection • Remote support communications with both audio and video and more • Auto-focus control • 8 megapixel camera • Built-in stereo speakers • Advanced Vuzix voice control
8	Body vital monitoring devices	• Monitors blood pressure • Heart rate monitoring • ECG capture • Blood glucose monitoring • SPO_2 monitoring

1.7   Industrial Internet of Things

Many industries, namely, chemical, energy, food and beverage, infrastructure, marine, mining, power and utilities, water and wastewater industries, can benefit from Internet of Things technology by monitoring the workflows to improve productivity and yield. It ensures safe and reliable operations. IoT helps actively drive value and sustainability to continuously improve manufacturing, provide more flexible and agile production capabilities, and optimize energy consumption and emissions. The Industrial IoT (IIoT) devices can monitor resources and idling times and help plan such that these dead times are reduced and productivity is increased. A typical IIoT solution is shown in Fig. 1.6.

Both IoT and IIoT devices consist of process/parameter capturing devices and control the surrounding subject of interest. The subject of interest could be us, processes, surroundings, and appliances which are controlled for their best performances.

1.8   Scope of Internet of Things (IoT) Technology

IoT and IIoT technologies have the potential of making everything smart and intelligent. "Thing" is used to mean anything controllable based on the conditions, automated, accessed anywhere and anytime. So, anything or any process which can

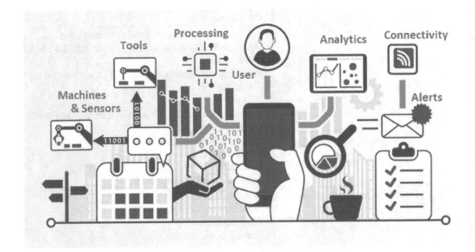

Fig. 1.6 IIoT system solution [1]

be adjusted by sensing it is suitable for automatic adjustment. The scope of IoT and IIoT depends on one's imagination, and hence, there is a huge scope for new inventions, innovations, and research. For example, things like windows, doors, chairs, tables, fans, lights, shoes, watches, and clothes can be made smart using sensors, actuators, and smart materials. Also, these devices need the Internet for accessing them anytime and anywhere to control them anytime. There are a lot of challenges to address while adopting these technologies onto these devices. Few challenges are focused on sensor and actuator technology that forms part of the devices, and others are communication technology challenges. IoT and IIoT devices use communication technologies to send/receive data sets or for further processing to analyze the information. Some of the challenges in devices are in signal capturing, processing, sensor and actuator designs. Main challenges in communication in these devices are in establishing reliable link with other devices through different media and accessing right data set in the most accurate and timely manner.

1.9   IoT Technologies

Different technologies involved in making IoT or IIoT systems or devices are listed below:

- Sensor technology
- Actuator technology
- Signal and data processing
- User interface and machine learning (UI and ML)
- Artificial intelligence
- M2M communication technology
- Internet technology
- Embedded system/PCB/system design
- Integrated development environment for system software development

The complexity of the IoT system depends on the application.

1.9.1   Sensor Technology

Most of the IoT and IIoT devices sense some physical parameters or capture an image or video or sound in some form. Sensors are specialized electronic devices which are used for this purpose. These devices capture physical parameters, which are processed further to extract the required information from the analog form. There are three types of sensors. The first type of sensors are pressure sensors, temperature sensors (thermistors, thermocouples, PRT), fuel gauge, etc.

These are the devices that sense and convert the physical parameters to electrical signals for processing and control. These are also called transducers. The second type of sensors are cameras, accelerometers, proximity sensors, GPS, etc. which give out the parameters in the digital form which can be processed directly. The third type of sensors are electrodes, which directly receive electrical signals from surroundings or contact surfaces, like ECG probes. Figure 1.7 shows some commercially available sensors. Sensor technologies are a crucial technology to realize IoT systems.

1.9.2 Actuator Technology

IoT devices or IIoT systems take action depending on the sensed parameter value. For example, in the home automation system, the room light turns on if it is dark and a person walks in. In this case, the sensor senses person entering/exiting the room and the light turned on/off. The light control is achieved by turning on or off the switch connected to the bulb. In IoT devices, the switch control is automatically done by the control circuit which activates the switch according to the sensor state. The switch control circuitry is called the *actuator* circuit. Similarly, there are other actuator circuits which control the valves and motors depending on the application. There are advanced actuator circuits which can even administer drugs

Fig. 1.7 Few commercially available sensors

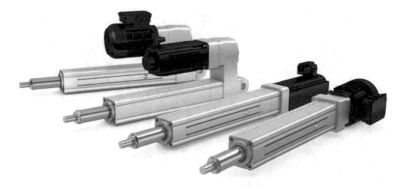

Fig. 1.8 IoT-ready electromechanical actuators [2] (Source: Ewellix)

into person's body depending on the sensed value of body vitals in health monitoring systems. Displaying the results with out-of-range alarm is also a simple actuator circuit. *Actuator technology* is another important technology to realize IoT systems depending on the application. Figure 1.8 shows some electromechanical actuators.

1.9.3 Signal and Data Processing

The sensed signal from a sensor is processed in special circuits called *signal conditioners*. Signal conditioning circuit provides proper bias, removes DC offsets, and averages or amplifies the sensor output signal thus improving the accuracy of the captured parameter. Typically, signal conditioning involves impedance matching, buffering or bridging, and amplifying the signals. In some applications, it is also called the *analog front-end* circuit which includes an analog to digital converter and serializer-de-serializer (SerDes). Once the captured and conditioned accurate output signal of the sensor is converted to a digital signal, it is processed in standard RISC processors or digital signal processors (DSP) depending on the application. Analog to digital conversion is done as it is easy to process digital signals. Figure 1.9 shows some of the well known processors where the data/digital signal processing is carried out in IoT/IIoT devices.

1.9.4 User Interface and Machine Learning

User access is a very important for the IoT devices for displaying processed information. Most of the IoT devices come pre-configured and do not require any major user input. However, an easy user interface is very much a requirement whenever the user has to coordinate with the device. Another promising

Fig. 1.9 Standard RISC/DSP processors used in IoT/IIoT devices

technology in IoT systems is *machine learning.* The IoT device configurations are fine-tuned by learning user preferences when user uses the devices. This is done by machine learning algorithms in the devices. For example, Sensing algorithms of the temperature and SpO_2 are adjusted learning the usage pattern in body vital monitoring device. The sensing method is adjusted suitably by learning the touch artifacts like orientation of the device, device holding pattern, pressure applied on the sensor etc.

1.9.5 Artificial Intelligence (AI)

Artificial intelligence (AI) is another technology used to make IoT smart. This is done by emulating human intelligence in the devices. The IoT uses AI technology by mimicking human thinking in algorithms to anlayse, the captured data and take intelligent decisions in the systems. This can happen when the device is capable of learning by observing the surrounding events and how they are addressed using sample reference data set. Data processing in smart IoT devices can be equipped with artificial intelligence for correct interpretation and statistical analysis of the data stored over a period. This can be further extrapolated with prediction algorithms to predict future conditions and assess potential risks and mitigate them using deep learning algorithms. *Deep learning* is one of the machine learning techniques used when there is a large amount of reference data set. Data set is a set of events and responses which have occurred in the past captured and stored in memory accessible to the device.

1.9.6 Machine to Machine (M2M) Communication Technology

Most sophisticated IoT network can be developed if IoT device is capable of communicating the data captured to a remote device for better consolidation or coordination. Typically, Machine to Machine (M2M) is the direct communication between two peer-to-peer (P2P) devices on a communication channel. The communication can be on standard wired or wireless medium. Advancement in communication technology over the years has made it possible to build a very reliable and robust network of IoT and IIoT devices. Depending on the applications, the network of IoTs can be customized architecturally for reconfigurability, scale-up or scale-down of the number of network nodes, low power, and more efficient coordination. More about M2M technology will be covered in detail in the subsequent chapters of this book.

1.9.7 Internet Technology

As in the name of IoT and IIoT, the devices use the Internet to communicate with other devices and applications on the Internet. Hence, Internet technology is the most important technology for making the things/processes smart. This enables access to the data captured and processed information by IoT and IIoT devices anywhere and anytime via the Internet. The Internet technology is the standard Ethernet technology based on IEEE802.3 standards [3], which will ensure that the device data set is accessible to stakeholders like users or applications for further consolidation, aggregation, processing, and storage. Information access on internet also enables new class of technologies called *Internet of software (IoS)* which will be discussed in more detail in further chapters in the book.

1.9.8 Embedded System/PCB/System Design

Major technology used in the design and development of IoT and IIoT devices is the embedded system based hardware platform. Typical embedded system comprises of printed circuit board (PCB), with processor modules and peripheral modules like display, battery, LEDs, electrodes, and sensors and actuators. The hardware platform depends on the requirements of the applications. The embedded systems also consist of software which is embedded in the flash or ROM memory in the processor VLSI chip and/or uses external memory like SDRAM/DDR for data storage. The architecture and design of the IoT using these technologies will be explained in the subsequent chapters in the book.

1.9.9 Integrated Development Environment for System Software Development

IoT and IIoT devices use inherently software running at different levels. Some of the software found in these devices are the following:

1. Boot loader
2. Embedded operating system
3. Custom embedded stack
4. M2M protocol software
5. UI and ML software
6. AI software
7. Internet software
8. Cloud software

1.10 Architectural Framework of IoT

IEEE 2413-2019 is the standard defined by IEEE for an architectural framework for the Internet of Things (IoT). An architecture framework description for the Internet of Things (IoT) [4, 5] which conforms to the international standard ISO/IEC/IEEE 42010:2011 is defined. The architecture framework description aims to address the concerns commonly shared by IoT system designers across multiple application domains like transportation, healthcare, smart grid, etc. for interoperability. Architectural framework for the IoT development includes descriptions of various IoT domains, definitions of IoT abstractions, and identification of commonalities between different IoT systems across domains. A conceptual basis for the notion of things in the IoT is provided, and the shared concerns as a collection of architecture viewpoints are elaborated in the body of the framework description. As it can be seen, IoT device, systems, or networks are cross-domain designs and implementations require product design, electronic sensors, computer science, and communication technologies. The standard is explained in the context of smart city which can be used as a reference model for arriving at architectural building blocks to design an integrated multi-tiered systems. The framework defines ways to document and mitigate architecture divergence for other applications. The requirements have been arrived considering the requirements from government regulatory and certification bodies, enterprises, and consumers, a variety of applications. Framework addresses the quality requirement that includes protection, security, privacy, and safety. The reference IoT model covers security requirements, energy efficiency in data communication, service requirements, and application aware routing. It helps identify design choices for IoT and match requirements within a specific domain structure to relevant design choice. Figure 1.10 shows different application domains and various stakeholders.

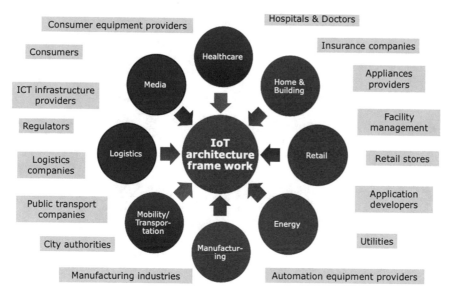

Fig. 1.10 IoT application domains and stakeholders

1.11   IoT and Open Systems Interconnection (OSI)

Systems are easily understood by representing them to standard OSI layers of communication. Figure 1.11 shows the standard OSI layered model.

As compared to conventional OSI layer architecture, in IoT layered architecture, layers 5, 6, and 7 are merged as application layer. Figure 1.12 shows IOT layered model as in OSI model.

1.12   M2M Communication Technologies

M2M (machine-to-machine) communication is the technology that enables IoT and IIoT devices to communicate with other devices. Most of the IoT and IIoT devices have M2M communication capability. This enables user to process more data, aggregate data sets from different devices, and get a bigger picture or derive larger perspective from the information gathered. M2M communication also enables users to:

- Gather valuable information about important processes
- Streamline and improve supply chains
- Predict costly maintenance and avoid mistakes
- Keep check on expensive and redundant assets
- Offer value-ads and/or services that would not be practical without connectivity to other devices like smart watch or smartphones

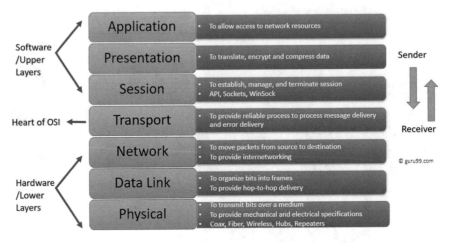

Fig. 1.11   OSI layer architecture of communication (Courtesy: guru99.com)

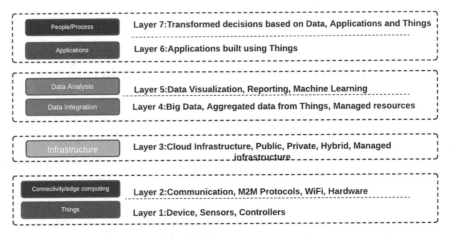

Fig. 1.12   IoT layer stack

M2M technology is often confused with IoT technology as most IoTs adopt this feature. M2M communication make IoT and IIoT devices inter-operable among them on network. Often IoT devices communicate to cloud server for high processing capability and storage of information. M2M technology can be wired or wireless and can have their own communication protocols. M2M technology is a point-to-point communication unlike IoT which could be via cloud. Following use cases are the examples of M2M technology with IoT and IIoT devices.

1.12.1 In-Vehicle Telemetry Services

Several car manufacturers offer in car connectivity services for their customers. Occasionally, these services allow people to use a built-in subscriber identification module (SIM) card to access the Internet on their smartphone or tablet. But, in-car connectivity is used to relay information about the conditions of different parts of the in real time while its usage to assess the health of parts to the manufacturer. Car Manufacturers use this data to improve future models and to also offer more hands-on help, contacting users if any part need replacement. This type of solutions allow car manufacturers to offer a better service to their customers.

1.12.2 Smart Meters

Smart meters (sometimes called utility meters) are the IoT systems with M2M communication for tracking electrical energy consumption in real time. They also allow the energy provider to observe the consumption pattern of the user. The energy provider can track power consumption and generate power in more efficient ways.

1.12.3 Smart Asset Tracking Services

Asset tracking is an important for many businesses, like in the shipping industry and/or businesses with a large fleet of vehicles. With recent advancement in M2M technology, businesses that need tracking their assets can now do so using relatively inexpensive global positioning system (GPS) trackers connected to an M2M network.

These GPS trackers allow the movement of vehicles to be tracked in real time. They also allow companies to gather useful data about the fuel consumption, average trip times, and driver performance parameters to improve the efficiency of journeys and processes.

1.12.4 Supply Chain Management (SCM) Solutions

Modern supply chains are very complex and incredibly large. M2M technologies, including smart barcode scanners, RFID systems, automatic stock management systems, and GPS tracking, help to keep these supply chains connected. They play a crucial role in allowing a business to track raw materials and stock and finished goods as they move through the manufacturing/retail process—providing companies with the data they need to refine their processes and remove the everyday roadblocks that reduce their profit and also redundant processes.

1.12.5   Wearable Technologies

Another vital group of the IoTs are wearables like smart watches and, Fitbit, which use of M2M technologies to send data about the physical activity, heart rate, and other important metrics to a smartphone, as well as receive texts, push notifications, and other communications about one's day-to-day activity.

Generally speaking, a smart watch or Fitbit will have a very limited processing power, which means that it relies on the smartphone's CPU to send notifications to the user when he/she has walked a certain number of steps or alert the user when it is time to head for some activity. In fact, these devices are primarily a gateway between a wrist-mounted sensor and the smartphone's computing power, and while they generally use short-range, low-bandwidth technologies to communicate.

References

1. Rajiv, "Applications of Industrial Internet of Things (IIoT)" RFpage.com, 23, June 2018
2. "New IoT-ready electro-mechanical actuators from Ewellix", Linear motiontips.com, 26-Nov-2019
3. IOT architecture from IEEE working group
4. IEEE standard -SA website for IEEE 2413 standard
5. IEEE 802.3 Internet standards

Chapter 2
Internet of Things (IoT) Design Methodology

2.1 Architecture Framework of the IoT System Solution

As IoT devices can be of large spectrum of complexity, sizes, and forms, they can be realized in so many different ways. And hence there are many possible architectures of IoT. Not all devices can fit into one standard architecture. However, considering following the common important features of IoT help to arrive at reference architectures:

1. Reconfigurability: It is necessary for the IoT device to be reconfigurable depending on the use case environment and network dynamics on the Internet.
2. Multiple sensor/actuator connectivity: IoT devices work with many sensors and actuators. In many scenarios, common the processing circuits are used to process the data captured by sensors. It is therefore essential for the IoT device to be able to connect to many sensors and actuators depending on the functional modes.
3. Scale-up/scale-down: The device architecture should have the flexibility to add more functions and modes as they are connected to a large number of sensors and actuators. Also, they should also have the necessary configuration to trim down the functionality based on the need and context.
4. Connectivity to the Internet or larger web: As the name indicates, the IoT devices are connected to the Internet or larger web for distributed functions and remote access.
5. Low power consumption: Low power consumption is very essential for IoT devices as they operate in the field and most times on battery or alternate sources like solar. It will be very difficult to visit the IoT sites to change the batteries or attend to related issues. These devices should have low power consumption for ease of maintenance.
6. Distributed functionality: IoT devices are connected to the Internet, and the data is processed in different system components like gateway and edge device or on cloud. Therefore, IoT systems are needed to support distributed computing.

V. S. Chakravarthi, *Internet of Things and M2M Communication Technologies*,
https://doi.org/10.1007/978-3-030-79272-5_2

7. Modular architecture: It is always preferred to have modular design for easy maintainability and provisioning the device on the Internet.
8. Interoperability: IoT devices are required to operate with a wide range of other IoT devices from multiple vendors and subsystems on the Internet. Interoperability is an essential feature to communicate with a wide range of devices on the Internet.
9. Tunable performance like latency, speed, power, and size: Since IoTs operate in very dynamically changing networking environment on the Internet, it is necessary for them to have performance adaptability in parameters like latency, speed, power, and size to get optimal reliable performance.
10. Flexibility in sharing the processing power from centralized to distributed: In the dynamic networking scenarios, there will be changes in the processing powers needed in different system components in the IoT systems. This requires balancing and distributing the processing load in different components of the system. It is necessary to build flexibility to configure the systems for centralized processing or distributed processing accordingly.

Though all the above features are essential for an IoT system, the criticality of each of them depends on actual requirements for a particular application. It makes it difficult to define one hard architecture for IoT-based system. However, it is possible to come out with the set of guidelines for defining the architecture which best suits the system requirement. The following types of architecture guidelines are suggested for an IoT system for any given application:

1. Three-layer architecture
2. Five-layer architecture
3. Cloud-based architecture
4. Fog-based architecture
5. Edge-based architecture
6. Mist-based architecture
7. Big data architecture
8. QOS-based architecture
9. Mobility first-based architecture
10. Cloud things-based architecture

Most of the applications fit into the first three architecture types, but a few IoT applications fit into edge-based architecture which is in demand. The following section describes three-layer architecture, five-layer architecture, and cloud-based architectures. Other architectures listed above are the result of system optimization targeted in distributed computing adopted in IoT-based systems. These can be considered depending on specific requirements of the applications.

2.2 Three-Layer Architecture

IoT system can be defined using *three-layer architecture* with layers defined below:

1. Thing or device layer
2. Network layer
3. Application layer

2.2.1 Thing or Device Layer

The thing or device layer consists of the actual devices with multiple sensors and actuators. This layer is responsible for interacting with the environment for sensing, capturing, and controlling the parameters or other things. These can be physical devices like cameras, sensors, electrodes, etc. The thing capture process and connects to Internet through wired or radio frequency (RF) links. Such devices or things are referred as *edge device*. The device function can be sense only or sense and connect to internet functionalities. So, while taking up the IoT design, it is essential to understand the requirements to be implemented in the device layer. Most IoT devices support M2M communications functionality through Bluetooth, Wi-Fi, or Ethernet. Nevertheless, the device features with sensors and actuators have to be customized for target application. The differentiation among things comes from the type of sensors/actuators the device support. Figure 2.1 shows the three-layer architecture of the IoT system.

2.2.2 Network Layer

In the IoT architecture, the devices or things access the Internet through gateways, switches, or routers. This layer is standard network layer optimized to collect and distribute data set generated. Network layer elements serve as messengers between the devices and the cloud layer or the Internet where actual application resides. The network layer functionality is implemented in devices or software programs that work with the edge sensors and other devices on the network. Large IoT systems might use a multitude of gateways to serve high volumes of edge nodes. They could provide a range of functionalities, but most importantly they normalize, connect, and transfer data between the physical device layer and the cloud server. Data transfer between the cloud server and the physical device goes through a gateway, sometimes called "intelligent gateways" or "control tiers." Many gateways support computing and peripheral functionality such as telemetry, multiple protocol translation, artificial intelligence, pre-processing and filtering massive raw sensor data sets, provisioning, and device management. It is becoming common practice to implement data encryption and security monitoring on the intelligent gateway to

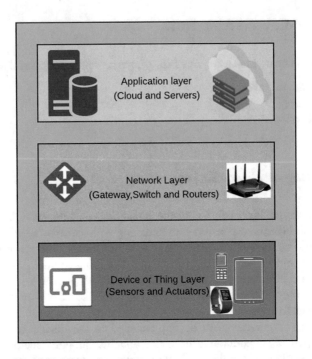

Fig. 2.1 Three-layer IoT architecture

prevent malicious "man-in-the-middle" attacks against otherwise vulnerable IoT systems. Certain gateways offer a real-time operating system specialized for use in embedded and IoT systems along with optimized low-level support for different hardware interfaces, with supporting interface libraries. Library functions for managing memory, input-output (I/O) timing, and interface are readily available to design network layer functionalities. Functional Libraries are generic functions based on standard protocols required for IoT development.

2.2.3   Application Layer

The cloud or data server is the application layer which communicates with the gateway, over wired or cellular Ethernet. It is powerful server systems with databases that enable robust IoT applications and integrate services such as data storage, big data processing, filtering, analytics, third-party application programming interfaces (APIs), business logic, alerts, monitoring, and user interfaces. In a three-layer IoT architecture, the "cloud" is also used to control, configure, and trigger events at the gateway and the edge devices.

2.3   Five-Layer Architecture

Another class of IoT architectures is the *five-layer architecture*. The first three layers -Device, Network and Application layers have same functionalities described in Section 2.2. Two additional layers in five-layer architecture are the following:

- IoT network management layer
- IoT data processing layer

Five-layer IoT architecture is shown in Fig. 2.2.

Fig. 2.2 Five-layer IoT architecture

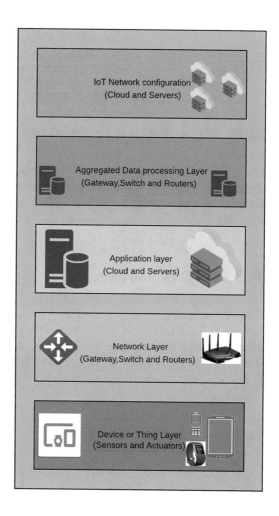

Table 2.1 IoT and functionality

Layer number	Layer name	Functionality
5	IoT network management layer	Manages IoT systems, dynamic changes of IoT network, and applications and collaborates with other cloud networks for bigger purposes
4	*Application layer*	Gives access to cloud for data storage, further processing of big data and data analytics. Runs user specific applications and interfaces with users
3	IoT data processing layer	Aggregates data from other sources and collaborates deep learning and big data processing
2	*Network layer*	Gateway or router converts device data to Internet data and connects the device on to the Internet. Functionality of this layer is to convert and transfer sensor data between layers and through network protocols such as LTE, LAN, Bluetooth, 6LPAN, LoRaWAN, etc.
1	*Thing or device layer*	Physical device which connects with a number of sensors and actuators. They collect data from the environment and control the parameters using actuators

2.3.1 IoT Network Management Layer

The IoT network management layer manages dynamic changes in the IoT network due to changes in the IoT environment, scale-up and scale-down of the devices, data paths, etc. to maintain consistent value to end users and other stakeholders.

2.3.2 IoT Data Processing Layer

The data processing layer is used in large IoT systems to process large and varied data sets like searching required dataset in big data set and optimize search engines of the database. Using the processing layer at more than one point within in a specific architecture may be required for a some IoT system.

Table 2.1 consolidates the functionality of layers in reverse chronology.

2.4 Fog-Based IoT Architecture

Applications like healthcare and vehicular automation need IoT architectures such that the data processing happens close to the device to reduce the latency. The *fog layer* lies in between the device layer and transport layer, i.e., gateways. This layer can be physical or logical in functionality. In fog architectures,

Table 2.2 Fog-based IoT architecture

Business layer	Manages the entire IoT system and its functionality, applications, and business models	
Application layer	Provides application-specific services to users	
Processing layer	Analyzes, stores, and processes large data sets. Might use databases, cloud computing, and big data processing resources	
Transport layer	Transfers sensor data between layers and through networks such as 3G, LAN, Bluetooth, LoRaWAN, etc. A typical IoT typical gateway	
Fog layer—smart IoT gateway	Security layer	Encrypts/decrypts data
	Storage layer	Stores files with local relevance
	Data processing	Filters, processes, analyzes, and reduces edge data or process commands or subscriptions from the cloud
	Monitoring	Monitors power, resources, responses, and services and access
Perception or physical layer	Sensors gather data from the environment. Actuators turn things on or off or set values	

computational and storage resources are provided typically in gateway devices or physical devices. When these resources are added to gateway device, it is called a "smart IoT gateway," and when these resources are added to the node, it is called "fog node." Layerwise functionalities of Fog layer IoT architecture is listed in Table 2.2.

2.5 Edge Computing Architecture

Whenever application demands real-time responses, sense and response latencies have to be very small. In such systems, data communication all the way to cloud server becomes expensive and slow. For such IoT systems, the captured data has to be processed close the device layer, and control function has to be activated in real time. Such an architecture is called *edge computing* architecture for IoT. The edge part of the cloud has to be equipped with resources for storage of real-time data and computing. The processing of user data (sense and response control functions) have to be done in the IoT device layer. Edge computing can be seen as fog computing, where real-time low-latency processing capabilities and functionality are performed closer to the edge nodes. It can be particularly useful in reducing sense and respond latencies for applications that require real-time performance. In edge computing, processing takes place at the physical device layer directly on a smart device connected to the device or on an IoT device itself. Edge computing decentralizes computing, increasing data privacy, and allows for mesh networking of the cloud.

Fig. 2.3 Cloud computing in IoT architecture

2.6   Hybrid Cloud-Fog-Edge Architecture

As stated in previous section, fog and edge architectures could be hybridized with cloud-centric IoT architectures, if deemed a good fit to meet a project's requirements. Figure 2.3 shows one combination which uses a nested configuration.

2.7   IoT Design Flow

IoT design flow involves several steps starting with the requirements and ending in validation as shown in the IoT design flow in Fig. 2.4.

2.7.1   Requirement Study

IoT development strategy and choice of technologies needed depend on a set of requirements. Study of the requirement for the IoT design depends on the problem definition. The requirement capture is done by asking set of relevant questions about the chosen problem. For example, if body temperature monitoring system has to be developed, it is required to ask the following questions:

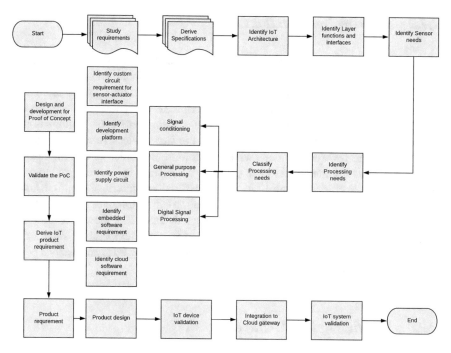

Fig. 2.4 IoT system design flow

1. Is the device noninvasive, hygienic, convenient, and affordable, and does it measure temperature that is close to the core body temperature?
2. Will the device be used in contactless mode or contact mode?
3. Will the device be used in self-monitoring mode or used with assistance?
4. What is the accuracy needed? Is there a standard to be met on monitoring body temperature?
5. What should be the maximum form factor of the device?
6. Will the device be battery operated or otherwise?
7. How many body temperature values the device should store?
8. Is it necessary to access temperature remotely and if yes what is the range of access needed?
9. Is it necessary to display the temperature in Celsius or Fahrenheit or both? Is audio output of the related messages required?
10. Is it necessary to process recorded temperatures over time?
11. Is there a similar device available in the market already? How can this product be different? (This will lead to business decision on why it is required to develop this device.) What is the unique selling proposition of this device in comparison to the available devices?
12. Should the device be handheld or tabletop?

The list of requirements is prioritized and used to derive specifications of IoT system.

2.7.2   Deriving Specifications

From the prioritized requirements, the product specification is derived. Deriving right specification is necessary as they dictate system component decisions. For the body temperature monitoring system requirement of contactless or contact-based temperature measurements, the answer determines the choice of the sensor. Selection of communication technology for IoT systems depend on combination of requirements. For example distance between the two devices (range) and power consumption are few requirements which determine the choice of communication technology to be implemented in IoT systems. If the range is less and need to work at low power consumption, than 10 m, it will be Bluetooth interface required, and if the range is 100 m and medium power consumption can be tolerated, Wi-Fi technology would be right choice technology is adopted. Table 2.3 shows the probable mapping of specification to the requirements for body temperature monitoring device. Please note that there can be many reasons including business reasons of the arriving at the specifications from requirements.

2.7.3   Identifying IoT Architecture

Once the specifications are frozen, it is necessary to decide on the architecture that is most suited. The layer architecture and the system architecture differ only in the location where the functionalities are implemented as shown in Fig. 2.5. Some parts of the bottom three layer functions shown in the right side of the figure can be implemented as the hardware-embedded software. The top two to three layer functions can be implemented in the generic systems or devices based on standard computing platforms. The two sections communicate with each other through communication system containing interfaces and channel.

For the body temperature monitoring system, under discussion, one can easily map all the system functionalities to the three-layer architecture.

2.7.4   Identify Layer Functions and Interfaces

From the specifications, components and platforms, Hardware -Software components and interconnections (interfaces) are decided to implement the the layer functions.

Table 2.3 Requirement to specification mapping

Sl. no.	Requirement	Answer	Specification	Rationale behind the mapping arrived
1.	Device has to be noninvasive, hygienic, convenient, and affordable, and also the measured temperature value should be close to the core body temperature	Device will be noninvasive, hygienic, convenient, and affordable, and the measured temperature value would be close to the core body temperature	Body temperature of a normal human being will be in range 36–38 °C. Therefore, the range of the device can be 20–50 °C	Healthy Human body has the body temperature in the range (37.2–37.8 °C), and increase in temperature beyond 41.5 °C or decrease below 30 °C is considered unhealthy
2	Will the device be used in contactless mode or contact mode?	Contactless	IoT device should support IR sensor-based body temperature measurement	Contactless temperature measurement is possible by IR sensors
3	Will the device be used in self-monitoring mode or used with assistance?	Self and with assistance	Device can be handheld	It is easy to support both self-monitoring and monitoring with assistance
4	What is the accuracy needed? Is there any standard requirement on monitoring body temperature?	0.5–1 °F 0.3–0.6 °C	Body temperature is one of the most important body vital parameters to assess health of the people; it has to be accurate to comply to medical standards	Proper functioning of the body is dependent upon keeping the body temperature within the normal range (37.2–37.8 °C), because an increase greater than 41.5 °C or decrease less than 30 °C is found to be unhealthy and dangerous. Any IR-based temperature sensor which can have accuracy of ±1 °C and digital interface for calibrating it is a good choice
5	What should be the maximum form factor of the device?	Comparable to available thermometer	It should be convenient to hold in hand	

(continued)

Table 2.3 (continued)

Sl. no.	Requirement	Answer	Specification	Rationale behind the mapping arrived
6	Will the device be battery operated or otherwise?	Battery operated	Power circuit has to be designed considering the battery as the power source. Product design should consider suitable battery case	The device has to be portable
7	How many body temperature values the device should store?	100	The temperature stored has to be tagged with timestamp to indicate when body temperature was recorded at bear minimum. To make the device smarter, the designer can also tag it with the user id. There should be provision to key in the user identity	Since the temperature measured could be of self or others, it is better to timestamp the measured value to give the user a hint to know whose temperature was monitored
8	Is it needed to access temperature values remotely and if yes what is the range of access needed?	Yes with a device at a distance of 100 m	Need Wi-Fi interface and support so that the measured data can be periodically sent to smart devices	
9	Do you need to display the temperature in Celsius or Fahrenheit or both? Do you need audio output of the related messages?	Yes; temperature value to be displayed in centigrade and Fahrenheit and also if possible, the value to be output as audio message	Need conversion and display support	Countries use both units of temperature
10	Do we need to process recorded temperatures over time?	No	No statistical processing needed	

(continued)

Table 2.3 (continued)

Sl. no.	Requirement	Answer	Specification	Rationale behind the mapping arrived
11	Is there similar device available in the market already? How can this be different? This will lead to business decision on why it is required to develop this device. What can this device have as unique selling proposition in comparison with the available device?	Yes. There are similar devices in the market	Cheaper, smaller, and longer range	These are diactated by user preferences
12	Temperature measurement device has to be convenient and easy to use	Yes	Input to product design	This feature can be unique if the design is aesthetically attractive

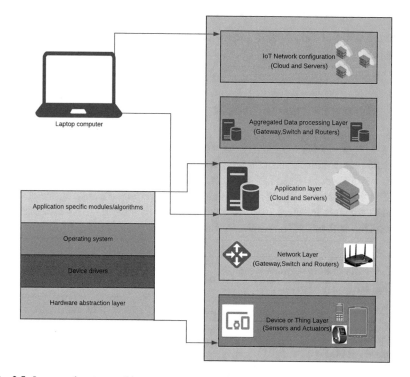

Fig. 2.5 Layer and system architecture

2.7.5   Identify Sensor Needs

The IoT system design for body temprature monitoring, needs to support IR temperature sensor and its interface circuitry along with calibration to maintain the accuracy and sensitivity level. IR sensors are more susceptible to temperature gradients, and hence careful choice of the sensor signal conditioning circuit is needed. There are many choices for sensors for this requirement. MLX90632 [1] is considered as a suitable sensor for this IoT system in this example design under consideration. MLX90632 sensor is an integrated circuit (IC) module with temperature interface block, programmable gain amplifier, ADC, and the digital interface to connect to the processor and internal interface to the on-chip memory like EPROM. It has the I2C slave interface for accessing internal function control registers. Sensor IC supports multiple power save modes, which can be used for reducing power consumption when not in use. The main features of MLX90632 are the following:

- Accurate and stable operation in thermally challenging environments
- $3 \times 3 \times 1$ mm quad flat no-lead (QFN) package
- Commercial grade: object temperatures between $-20\,°C$ and $200\,°C$ with typical accuracy of $\pm 1\,°$ C. Medical grade: object temperatures between $-20\,°C$ and $100\,°C$ with an accuracy of $\pm 0.2\,°$ C in the human body temperature range

More details of MLX90632 sensor IC can be found in the data sheet of the sensor [1].

2.7.6   Identify Processor Needs

Since the requirement is to display the temperature, conversion of temperature units (C/F), and remote communication with Wi-Fi, a suitable processor and its peripherals have to be selected. In the IoT thermometer example, MLX90632 is a complete module with all necessary circuitry for the identified specifications, and it needs simple processor for user interface and communication. Any simple MCU board such as Arduino Uno will suit the requirement. Table 2.4 lists major features of Arduino Uno.

Table 2.4 Main features of Arduino Uno

Operating voltage	5 V
Input voltage	7–12 V
Digital I/O	14 (of which 6 provide PWM output)
Analog I/O	6
DC current per I/O Pin	20 mA
DC current for 3.3 V	50 mA
Flash memory	32 KB (ATmega328P) of which 0.5 KB used for boot loader
SRAM	2 KB (ATmega328P)
Clock speed	16 MHz
LED built in	13

More details of Arduino Uno can be found in the datasheet. One can also go for faster MCU-based boards like STM32, which have better performance and features if future enhancements are considered. The STM32 board has a 32-bit ARM Cortex M0 processor which is Arduino integrated development environment (IDE) compatible. The communication interface can still be simple I2C interface. Few of the main features of STM32 in comparison to Arduino Uno are listed in Table 2.5. These boards need embedded software corresponding to the modules shown in Fig. 2.6 from MCU library. Embedded software development on this hardware is done using Arduino integrated development environment (IDE) on any windows or UNIX systems. For programming the board with the boot loader, the user guide from the board vendor or videos can be referred. A flash programmer has to be connected to the board and computer on which IDE is run. More detailed procedures with explanation are given in the subsequent chapters.

Table 2.5 STM main features

Hardware development board	STM32F103C	Arduino Uno
MCU	ARM Cortex M0	ATmega
Size of data bus	32	8
CPU frequency	72Mhz	16Mhz
RAM	20 KB	2 KB
Flash	64 KB	32 KB
GPIO pins	37	14
Interfaces	SPI, I2C, UART, CAN	SPI, I2C, UART
ADC pins	10	6

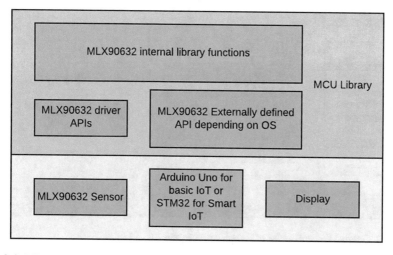

Fig. 2.6 IoT system architecture

2.7.7 Classify Processor Needs

The processor module should have enough processing power to load the measured body temperatures, store them to flash memory, and display them on a simple LCD display connected to it. Arduino Uno MCU has a processing power of 16 MIPS (million instructions per second) which should be good enough to perform the desired processing of temperature monitoring, calibration, and reading stored multiple temperature data set. Even STM32 boards which have better performance in terms of speed with respect to Arduino Uno can also be used for the application for faster performance like low latency and response time. It is assumed that the Wi-Fi module has its own processor which will take care of the communication of the data on Wi-Fi interface connected to digital interface. There are IoT development platforms with integrated Wi-Fi and Bluetooth for communication. Examples of such development boards are ESP32 series (which has Xtensa 32 processor) and Arduino variants. More information on this is available on 32-bit RISC-V MCU and 2.4 GHz Wi-Fi and Bluetooth LE 5.0. These offer hundreds of MIPS processing power and contain multiple RISC processors supporting complete Bluetooth and Wi-Fi protocols. More about these M2M communication is dealt in Chapters 10, 11, 12. These are complex development platform based on complex system on chip (SoC) which has more features like security functions, Ethernet, sensor sets, and interfaces. Figure 2.7 shows the block diagram of ESP32-like system.

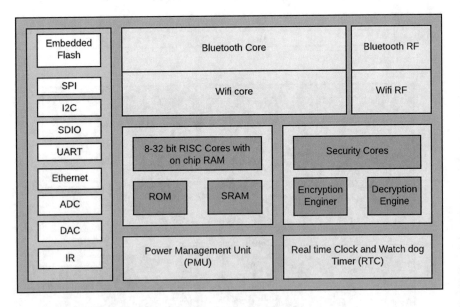

Fig. 2.7 Internal block diagram of complex development board

2.7.8 Proof-of-Concept Development

Complex IoT systems, are developed and validated on standard development platform or on the breadboard for proof-of-concept (PoC). For simple IoT systems as in body temperature monitoring device, it is not necessary to develop PoC but can directly be taken up for product development. Generally, PoC system is developed using standard development platform of selected micro controller unit (MCU). Typically, all processor chip companies have development boards with all possible interface to demonstrate the processor performance. This is used to develop any target application to validate the idea and study the feasibility. The PoC of the product idea involves development of the software stack needed for the particular application on a standard development platform and validate it in real application scenario. Standard integrated development environment (IDE) for the chosen hardware development board is used to develop the embedded software. The PoC development and the prototype development process will depend on the *software development environment* (*SDE*). It is necessary to debug and fix as many software issues as possible during development, and the quality of the software is determined by SDE and debugging capabilities of the developer. An *integrated development environment* (*IDE*) provides a graphical user interface for the tools in the software development environment. Different modules of IDE are compiler, debugger, flash programmer, and test simulation modules with the graphical user interface. The PoC hardware development platform includes the following on the PCB:

- The processor or, in the case of high-performance systems, several processors. Some of the processors may provide only limited programmability, as is the case for many video accelerators.
- The set of I/O devices for external interfaces.
- The standard bus interface.
- The software development environment.

The PoC software development environment (SDE) includes:

- A hardware abstraction layer (HAL), board support package (BSP), or basic input/output system (BIOS)
- Device drivers as standard APIs
- Operating system
- Hex or binary conversion of the program to program the on-chip flash memory in the MCU

Figure 2.8 shows software stack of the embedded PoC system. The hardware platform is at the bottom layer. The hardware abstraction layer provides basic software functions called *boot loader*. Software drivers may work through HAL functions or directly on the software. The real-time operating system controls the execution sequence of functional tasks or schedules them to give the user an impression of virtual concurrency.

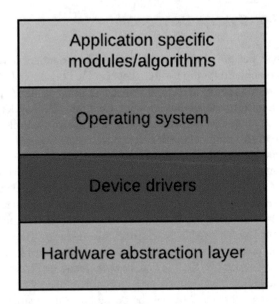

Fig. 2.8 Layer diagram for the embedded system

While the hardware abstraction layer may come with some I/O signal drivers for standard interface functions such as USB, SPI, the designer may need to develop these drivers if they are not available. The details of driver design depend on the type of operating system or executive used. A real-time operating system is designed specifically to provide real-time responsiveness for both I/O and process execution. Linux is widely used in embedded devices; not all versions of Linux, however, provide highly responsive I/O and real-time operation. The top-level software units are generally tasks or functions, each of which is a single thread of execution guaranteed to terminate in a finite amount of time. A task or function may run more than once, either sporadically or periodically. A complete application may be composed of several tasks and functions.

Sometimes the custom interfaces and application-specific board-like sensor and signal conditioning circuit, special power supply circuit, etc. will be developed as an add-on board depending on the need, which are interfaced with the development board.

2.7.9   Validation of PoC

PoC system developed for complex IoT system is validated for the intended function as per the specification. This is done in three stages: during software development, using test cases with the application programming interface (API)s available in the IDE, emulation in special emulation platforms (SEP), and "in-circuit emulation platforms" which needs extra hardware to be developed and interfaced with the PoC system to test the concept in NEAR real application scenario.

2.7.10   Derive IoT Product Requirement

Once the system is validated as a PoC system, actual product design and development can be taken up with full confidence. Most of the times, the complete PoC software stack will be reused for the actual IoT system. Product development activity starts again from deriving finer specifications from PoC both electrical and mechanical (product) like form factor, aesthetics, package material choice, etc.

2.7.11   Derive IoT Design Requirement

The device consists of processor subsystem, sensor interface, and power circuitry to be elaborated as design document which will be used for PCB schematic design. PCBs are to be designed and fabricated. This requirement is used to arrive at the list of components required for circuit diagram as bill of materials. These components have to be sourced from suppliers for assembly on the PCB.

2.7.12   Product Design

Product design involves package design for housing the electronics and the user interface like keypads and displays. The material for box design, safety and radio emissions, module positions like antenna positions, aesthetics, ease of use, and look and feel of the product are considered for the product design. This is the most creative part of the development which gives the first look to the user.

2.7.13   Product Validation

Once the product is assembled, it has to be validated for intended scenarios and use cases. On successful completion, the product is ready for marketing.

2.7.14   Integration to Cloud Gateway

Real field validation to check remote connectivity with interoperability is targeted at this stage. This may require special communication software application at the product level. This involves all the system components and subsystems in real case scenario as in the application scenario.

2.7.15   IoT System Validation

This involves regression testing for compatibility with previous similar devices and validation with necessary safety, regulatory requirement, and compliance tests submitted with real data to apply for regulatory certifications. Since most IoT smart devices are consumer devices, they are expected to get regulatory certifications for marketing.

2.8   Choice of Technologies

IoT system design requires many technologies as discussed in Chapter 1. Hence, choice of right technologies is very crucial for the success of the project. Development of IoT depends on the requirement, availability of the hardware development platform, sensors and associated circuit, power supply circuits, interfaces, and embedded software development platforms.

2.8.1   Hardware Development Platforms

Once the processor is identified, the IoT proof of concept (POC) is implemented on the development platform for the chosen processor. These are computing platforms with embedded software. Using the development platform, it is possible to develop part of the system or complete system applications which can perform computation and/or communication functionality in IoT system. These platforms are hardware systems with microcontroller unit (MCU). There are many MCUs, developed by companies like Atmel, Microchip, Texas Instruments, and Intel. MCUs are available as single-chip module with many peripheral functions like flash memory, ADC, DAC, on-chip RAM, EEPROMS, and interfaces like UART, I2C, SPI, LCD interface, etc. They operate at clock speeds of 16 MHz, 32 MHz, or more and perform data transfers on 8- or 16-bit bus. For example, Arduino boards are popular development board with Atmel's AVR MCU chip. Intel's

Galileo board is based on Intel's Pentium architecture. Intel's Edison is based on 500 MHz Intel's ATOM X86 CPU. Raspberry Pi is a development board based on ARM Cortex multiple cores.

2.8.2   Embedded Software Development: IDE

The software that is embedded in the hardware platform for stand-alone operation of the system is called embedded software. Embedded software is stored in flash memory of system on chip or on-chip memory like EPROM in the processor module. This software will be:

- Boot loader
- Embedded application stack consisting of

  - Device driver
  - Event handlers
  - Initialization software for communication modules, Analog front end (AFE)/ sensor module, configuration/mode setting, etc.
  - Software to work with hardware accelerators
  - User interface applications

The embedded software is developed on an integrated development platform (IDE) that will have software compilers, debuggers, and flash programmers' modules integrated in it. IDE platform will be dedicated to the hardware system as this software is targeted to be embedded into processor subsystem, for example, the Mbed software for ARM microcontrollers.

2.8.3   Custom Hardware Development

The development platform has most of the standard interfaces and some common functions. But many times, hardware development boards will not be enough for implementing all the functionality of chosen IoT application. Hence, there is a need to develop application-specific functionalities as an add-on board called daughter boards and interface them to the standard development board as shown in Fig. 2.9. The add-on board can have a specific sensor interface circuitry like signal conditioning circuits, analog front-end or SerDes circuits, or specific electrode/touch key interfaces. They are interfaced to the main board through GPIOs or through standard interfaces.

Fig. 2.9 Application-
specific daughter board
and standard development
platform for PoC system

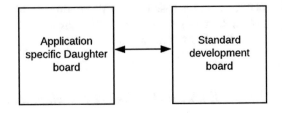

2.8.4   Product Design

Once the PoC system is fully validated, the product requirements are reviewed, and
final product specification is derived. In final product development or prototyping,
in addition to targeted functionality, design should address user safety requirements, environment safety requirements, interoperability, and co-existence with
other technologies with which this product operates.

2.9   Standard Compliance

It is necessary that all devices are to be safe to user and environment. The
International Electrotechnical Commission (IEC) [2] prepares and dictates the
safety and environment and safety standards for all electrical, electronics, and
related products. IEC works closely with IEEE and ITU who define functional standards for various technologies to define safety needs. All products are required to
comply to relevant applicable standards. The list of the standards available can be
referred on their respective websites [2]. There are test houses which test and certify
functionalities and adherences to these standards. For example, Wi-Fi part of a system has to comply to IEEE WLAN standard requirement. Similarly, Bluetooth functionality has to comply with Bluetooth standard defined by Bluetooth SIG. This is
for interoperability and co-existence of technologies which work in similar space.
There are some special standard compliances needed for automotive devices and
systems as well. List of few standards and bodies which define them are listed in
Table 2.6.

2.10   Regulatory Requirements

Apart from the standard compliance requirement, for most devices, there are
regional and global regulations to be followed. This will protect the safety, security,
and privacy of the users of these technologies. To develop and market the product in
a region, it is essential to safeguard the interest of the users of that region, and hence

Table 2.6 List of few standard bodies and standards defined by them

Sl. no.	Standard body	Standards
1	Internet Engineering Task Force (IETF)	Open international community of network designers, operators, vendors, and researchers concerned with the evolution of the Internet architecture and the smooth operation of the Internet Develops high-quality technical standards by working with related standard bodies and recognizes and enables cross-functioning of standard bodies
2	International Telecommunication Union (ITU-T)	Coordinates standardization work among technical study groups (SGs) in which representatives of the ITU-T membership develop standards for the various fields of international telecommunications and information communication technologies like cybersecurity, machine learning, etc.
3	European Telecommunications Standards Institute (ETSI)	Aims to develop open inclusive and collaborative environment for timely development of applicable standards in emerging technologies including IoT
4	Open Interconnect Consortium (OIC)	Defines mandatory core, optional, and security framework for interoperability across technologies
5	One M2M	
6	IEEE 11073™	IEEE Standard for Health Informatics—Point of Care Medical Device Communication
7	IEEE 2030™	IEEE Guide for Smart Grid Interoperability of Energy Technology and Information Technology Operation with the Electric Power System (EPS), End-Use Applications, and Loads
8	IEEE 1901™	IEEE Standard for Broadband over Power Line Networks: Medium Access Control and Physical Layer Specifications
9	IEEE 802.15.4™	IEEE Standard for Local and Metropolitan Area Networks Low-rate wireless personal area networks (LR-WPANs)
10	Open Connectivity Foundation (OCF)	Aims to develop specifications, framework, code, and a certification program to enable manufacturers to bring OCF-certified products to the market

the certification is required to market the product in that region. The European Commission, CE, CDSCO, and FDA are few of the known regulatory bodies which certify the product to ensure user safety and privacy requirements (Table 2.7).

2.11 IoT Device Development Platforms

There are many development platforms available on which the IoT system can be developed. Choice of the development platform is critical to the success of the IoT device. Embedded and desktop development platforms come with pre-built toolchains, debuggers, uploaders, and frameworks which work on popular operating systems such as macOS, Linux (+ARM), and Windows. The embedded software development board

Table 2.7 List of regulatory bodies and domeins addressed by their standards

Sl. no.	Regulatory body	Sector
1	TRAI: Telecom Regulatory Authority of India	Telecommunication and tariffs and cybersecurity
2	FSSAI: Food Safety and Standards Authority of India	Food, safety
3	EEPC: Engineering Exports Promotion Council of India	Trade and investment
4	CDSCO: Central Drugs Standard Control Organisation	Medical devices and drugs
5	EU: European Union— Medicines and Healthcare products Regulatory Agency (MHRA)	Medical devices including IoT
6	FDA: Food and Drug Administration	Regulates foods, drugs, biologics including vaccines, medical devices, electronic devices that give off radiation, cosmetics, veterinary products, tobacco products
7	FCC: Federal Communications Commission	An independent agency created to regulate communications by radio, television, wire, satellite, and cable, sought to consult citizens on proposed changes to rules impacting net neutrality

depends on the choice of processor and the interfaces required to connect with the external environment. Main consideration for this choice is the interface to software development environment (SDE) with the computer connectivity, interfaces to interact with environment which depends on the application. Most platforms come with communication interfaces like Ethernet, USB, UART, SDIO, JTAG, and GPIOs which are digital iOS and few analog iOS for analog functionalities. IoT development platforms are available as online development environment which provide large storage for developing large code base remotely.

2.12   IoT Device Data Management Platforms

When a large number of IoT devices are networked, they generate enormous amount of data which needs to be stored, processed, analyzed, and visualized in a manner that makes sense for use. There are a number of development platforms which have different modules integrated and available for development of database storage, analyzed and management functions. Some of the development platforms are listed in Table 2.8.

Table 2.8 IoT development platforms

Sl. no.	IoT development platforms	Description	Vendor
1	IBM Watson	Remote development platform for secure IoT data transmission, strong remote device control, and cloud capacity for storing large data, security features, and real-time data assessment with efficient risk management. Complete connectivity, data handling, and real-time data analysis. The real-time data received from the connected devices is handled at many points and is also organized and coordinated using data services. Due to secure design, this platform offers designers an efficient method to ensure completeness of IoT solutions	IBM
2	**Microsoft Azure IoT**	The Azure IoT is an extensive collection of Microsoft-related cloud services which connect and handle many IoT assets. You can develop and secure more IoT applications. Azure IoT can control any type of device, tool, security feature, and data analytics to meet the IoT objectives. This platform incorporates identity registry, device shadowing, data monitoring, and a set of the rule engines. Azure IoT Suite integrates with Azure Stream Analytics for processing a large amount of data created by the sensors	Microsoft
3	Google Cloud	This is a comprehensive IoT app for development and handling of IoT devices connected across the globe. It features an advanced analytics tool that allows companies to receive intuitiveness in real time. Google Cloud also offers combined services with cloud, end-to-end security, advanced data analytics, business process optimization, and a completely managed infrastructure	Google
4	**Amazon Web Services (AWS)**	It is a popular IoT platform available these days. It provides an exclusively solid and efficient framework platform in the cloud. IoT Device Management of Amazon Web Services allows easy extension and connection of devices. The administration assures a versatile and safe application with results of investigation, monitoring, and refreshing the device's utility	Amazon Inc.
5	**Cisco IoT Cloud Connect**	This IoT development platform provides developers secure and easy solutions for IoT along with other purposes like data analytics, network connectivity, app enablement, automation, and management	Cisco
6	**Arduino**	Arduino is an easy-to-use open-source IoT development platform. It offers a wide range of hardware specs to associated devices. Arduino's software comes with the strategy of the IDE (integrated development environment) and Arduino programming language	Arduino community

(continued)

Table 2.8 (continued)

Sl. no.	IoT development platforms	Description	Vendor
7	**Oracle IoT**	Oracle IoT platform supports programming with the range of devices and their analysis Oracle supports the creation of extremely large amounts of data, for development of IoT systems. State-of-the-art security elements that protect the IoT frameworks from outside risks	Oracle
8	**ThingsBoard**	This platform is for device management, data collection, visualization, and processing. It supports all incredible IoT protocols like HTTP, MQTT, and CoAP as fast as cloud and on-premise distributions. It creates workflows depending on RPC requests, REST API events, and design life cycle events	ThingsBoard Inc.
9	**Qualcomm's IoT Development Kit**	It is a blend of software and hardware. This IoT platform has incredible home automation and environmental monitoring	Qualcomm
10	**Kinoma**	This Marvell Semiconductor hardware prototyping platform includes three different open-source platforms. Kinoma Create is a DIY development kit for prototyping gadgets Kinoma Studio is the construction environment that works with setup and the Kinoma Platform Runtime. Kinoma Connect is a free Android and iOS app that connects smartphone devices and supports IoT devices	Marvel Semiconductor

2.13 Product-Hardware-Software Partitioning

Development of IoT product is a multi-domain activity. The functionality of the IoT device is spread across technology domains for simple and easy user interface and hardware where the actual electronic circuitry resides and the embedded software in on-chip flash memory or on-board RAM memory. The sensors and electrodes which have to interact with the user or environment will be fitted on the product case. The electronics hardware will be on a PCB in the product case along with display, keypad, keyboard, and other peripherals for user/environment interface. The device drivers and other software stack are embedded in the memory of the hardware unit. Depending on the use cases, the product may be designed for aesthetics. Figure 2.10 shows a complexity involved in the case of wearable watch IoT design with its internal parts.

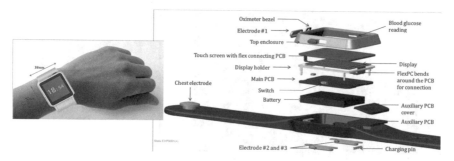

Fig. 2.10 Wearable watch and design exploration with internal parts. (Courtesy: SenseSemi Technologies Pvt. Ltd.)

2.14 IoT Software Development

IoT device development involves hardware and software. The IoT software development is carried out in the software development environment (SDE) of the targeted hardware development board. SDE is also called integrated development environment which is used for embedded system software development. It supports multiple programming languages like C, C++ Embedded C, system C, etc. It will have built-in compiler, simulator, flash programmer, and debugger running on the personal computer which can be connected to the development board through programming cable dedicated to the particular board. Some software development platforms run on cloud server which can be accessed on the Internet. One can develop the entire IoT software on the cloud server and generate executable bit code which can be downloaded and programmed onto the board through the local computer system through the programming cable. Embedded software development needs the following software modules:

- Operating systems (Windows CE, Yocto Linux, ThreadX, Nucleus RTOS)
- Languages (C, C++, Python, JavaScript, etc.)
- Tools (IDE, PDK, SDK, compiler toolchains, hardware and software debuggers (e.g., ST-Link, Segger))

Figure 2.11 shows the embedded development environment.

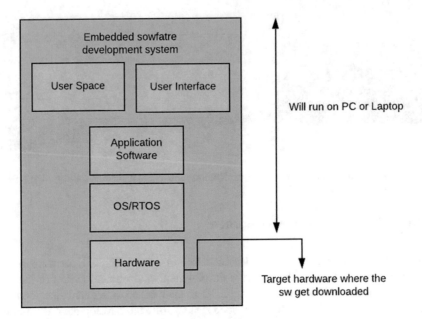

Fig. 2.11 Embedded development environment

2.15 Communication Technologies

The IoT devices usually have internet access by itself or through machine-to-machine (M2M) communication to the nearby smart devices which support Internet access. The M2M communication is achieved by either wired cable or wireless technologies like Bluetooth, wireless LAN or Zigbee, or LoRaWAN or 6LowPAN. More on these technologies will be dealt in detail in the subsequent chapters.

References

1. MLX90632 FIR sensor datasheet
2. www.iec.org

Chapter 3
Industrial IoT (IIoT) Design Methodology

3.1 Introduction to Industrial IoT

IoT products are used in consumer applications, like smart home, smart parking, and smart appliances, but also in automation of processes in industry and manufacturing. IoT solutions used in process automation in oil and gas industry, chemical industry, and product manufacturing require close process monitoring and control. Application of IoT in industrial automation called fourth industrial revolution I4.0. IIoT reaps the benefits of connecting machines and industrial equipments to a secure network that enables automation of the entire industrial facility. IIoT solution uses a combination of M2M communication, big data analytics, supervisory control and data acquisition (SCADA), and similar industrial equipment. This enables monitoring of local and remote industrial processes in real time for better informed quality decisions and optimization, leading to higher efficiency and productivity. Advantages of industrial automation or process automation in manufacturing are the following:

- Increased productivity
- Optimized process/workflows
- Reduced downtime
- Better inventory management
- Increased security
- Remote control
- Increased return on investment (RoI)

Few examples of IIoT-based automation are asset management, safety operations, and predictive analytics for maintenance with minimum downtime. In such applications, the conditions of the machines can be learnt continuously to plan the preventive maintenance so that the production is not hampered.

© The Author(s), under exclusive license to Springer Nature Switzerland AG 2021
V. S. Chakravarthi, *Internet of Things and M2M Communication Technologies*,
https://doi.org/10.1007/978-3-030-79272-5_3

3.2   Challenges of Industrial IoT

There are many challenges to adopt IIoT in manufacturing sectors. They are as follows.

3.2.1   *Heterogeneous Process Flows and Machineries*

A smart manufacturing unit will need to have a mechanism through IIoT devices that collect, integrate, merge and combine data from one machine to another. This requires robust industry-grade sensors in smart IIoT devices or compatible machines and equipments. The manufacturing Industries use different process flow and machines. For example, the food processing flow is different from paper making. VLSI fabrication process is different from PCB assembly flow. Sensing the process parameters needs a large number of sensors to capture meaningful process data from these machines. Connecting different types of process machines is challenging. Communication technologies like such as Bluetooth and Zigbee have limitations in aggregating and communicating process data in automating industries as there will be large number of sensor nodes. Developing industry-grade sensors requires extensive training and simulation within the ecosystem, which can prove difficult, and may need large investments. Such sensor based devices and mechanisms are not yet available for most manufacturing units. And developing industry-grade sensors require special skilled workforce and fabrication facilities within the ecosystem, which can prove difficult, and may need large investments.

3.2.2   *Process Security*

Autonomous manufacturing is prone to security threats. Any process hack can cause huge damage to the units resulting in losses for both the business and the employees on site. This vulnerability may cause malfunction of the manufacturing process, potentially putting the safety of employees and machines at risk. These have to be addressed by making the machines more robust and secure from cyber threats.

3.2.3   *IIoT-Based Automation*

IIoT adaption in manufacturing is still at early stage, being achieved to a great extent by programmable logic controllers (PLCs), single board controllers (SBC) and remote terminal unit (RTU) and a few machine automation sensors. The SCADA software collects of data from machines, process and display them to the responsible person responsible to take early decision regarding process flows. For example, the automatic process control system can notify large deviation in process parameter to the user to tweak the process control parameters after analyzing the root cause. This will prevent further loss of the products but cannot undo the loss incurred.

3.2.4   Oil and Gas Automation

IIoT use to automate oil and gas processing plants offers altogether different challenges. These plants are high-risk industry with extreme operating conditions, often involving highly combustible substances and expensive equipments. Main challenge is that of remote desert location and the expensive infrastructure. IIoT solutions have to be much more rugged and robust to monitor in extreme weather conditions and should have fast control reactions. The system should schedule maintenance to minimize sudden breakdowns.

3.3   Automation Framework Using IIoT System Solution

Figure 3.1 shows the industry automation framework using IIoT systems with SCADA system. The SCADA system will perform the following functions:

- Data acquisition
- Data communication
- Data reporting, representation, and visualization
- Monitoring and control

The SCADA system consists of a software program to provide trends and diagnostic data and manage information such as scheduled maintenance procedures, logics information, detailed schematics for a particular IIoT device or machine, and expert system troubleshooting guides. This process operator can see a schematic representation of the plant being controlled. It has a set of IIoT devices collecting various machine and process data, which are connected to one or more interconnected programmable logic controllers (PLCs) or remote terminal units (RTUs) of the SCADA system which are further connected to data servers which have manual control for PLCs/RTUs and human-machine interface (HMI)-based displays. This architecture is implemented in many ways. The IIoT and PLCs can be connected by M2M communication technologies and one of the single-board computers (SMC) with M2M node to the Internet, which in turn links to data centers, local or remote. The data server has human-machine interface through which the process is controlled. The topology of the network architecture is application dependent. IIoT automation system can be configured as clients and server, where multiple process

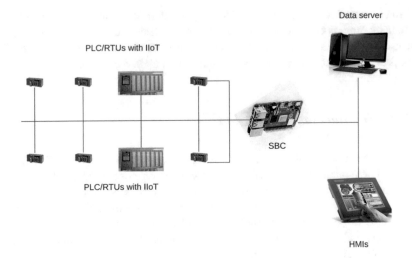

Fig. 3.1 SCADA system components

sensors on the machines and the switch/relay controls using PLCs are the clients on a typical data server where large data set analyzed to control the manufacturing process.

Different components of IIoT system in SCADA network is explained below:

3.3.1 PLCs/RTUs with IIoT Devices

The process or machine sensors in manufacturing plant monitor flow, pressure, temperature, etc. of the process, which are controlled. IIoT devices collect the process parameters from the machines and transfer them to a programmable logic controller (PLC) or remote terminal unit (RTU). (PLC with M2M communication capability is called remote terminal unit (RTU).) RTU device may have direct Internet connectivity in the case of small manufacturing unit or M2M communication to gateway device to aggregate and send the processed data to a nearby SBC or directly to the data center on the Internet for remote monitoring. The communication interface can be wired or wireless. Technologies like synchronous optical network (SONET) or synchronous digital hierarchy (SDH) are well-known communication protocols used for this purpose. SONET is a synchronous data communication protocol used to transmit large data on optical fiber line, and a variant of SONET is SDH which supports higher-order multiplexing for large synchronous or asynchronous data transfers.

3.3.2   Single-Board Computers (SBC)

Single-board computer (SBC) system aggregates the data collected from different IIoT devices and PLCs/RTUs and analyzes them before transferring to data servers, on the Internet. SBC connects with PLCs via M2M interface to collect the process data periodically or on request. On the other side, they can transmit the processed and analyzed data on the Internet to the data server.

3.3.3   Data Servers

Data servers store the process parameters as a database; perform search, statistical, and performance analysis on a large number of stored historical data (big data processing); and provide decision support to the process in-charge. Data center has human-machine interface where the person in charge of the process will access real-time process data and decision support system to make informed decision to change the PLC configurations and control the process machine settings for fine-tuning the process if required. Process in-charge can also configure through these systems, suitable alerts, and alarms for process controls. These systems can even send text and voice alerts to the concerned persons by SMS on real time. Based on the history of how "out-of-range data set" was handled in the past, it can even recommend an action to the process in-charge.

3.3.4   Human Machine Interface (HMI)
and Supervisory System

HMI is the master control system which is networked with data server and many machines depending on the type of industry. User can interact, generate reports, and modify the configurations of the system setup, PLCs, and IIoT devices to operate relays or switches to control the process parameters. This system will be fault tolerant with good backup as it is the main control of the SCADA system. Communication link of the supervisory system in SCADA network has to be very reliable for this, especially in applications like railway system control or power generation plants. They can use SONET and SDH for reliability and performance. The network visualization is a very important factor of the IIoT networks, as it has to be very user friendly. The status, error, and action information of different components of the network will be shown on process maps with figures and physical coordination as many times, control action has to be taken manually on site.

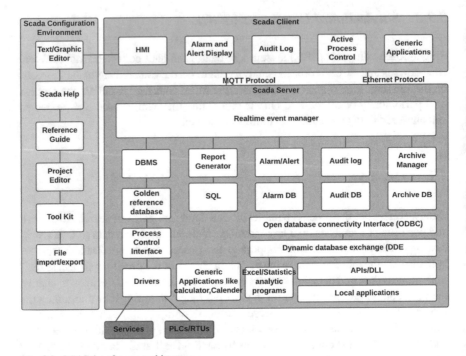

Fig. 3.2 SCADA software architecture

3.3.5 SCADA Software

SCADA software architecture for manufacturing is shown in Fig. 3.2. It consists of functions for managing big data, process control and monitoring involving user interface (human-machine interface), progress/diagnosis, report generation, data analytics, alert and alarm system, command audit log database (database of commands executed by the user) and archive database, and also actual schematic visualization of the entire plant with workflow and status of the processes live.

3.4 Technologies in IIoT Solution

IIoT concept is not new in industry automation. It existed as embedded solutions in earlier days. These machines are becoming smarter day by day due to the advancement in machine learning, artificial intelligence, big data processing, and smart user interfaces apart from developments in sensor/actuator technologies. Major technologies contributing to the industry automation in recent days driving I4.0 revolution are the following:

3.4.1 Sensor and Actuator Technology

In the manufacturing domain, the I4.0 vision has promoted smart manufacturing and smart factories concepts by augmenting all assets with sensor-based connectivity. Sensors connect humans, machines, processes, and systems in manufacturing environment. Intelligent sensors such as positioning tags, safety gloves [1], head-mounted displays (HMDs), man and machine body temperature monitors, and smart glasses [2] have been widely applied in industrial applications. RFID technologies and sensor network technologies enable tagging, tracking, and monitoring of personnel controlling the machines and machines themselves. These intelligent sensors generate a large volume of industrial data which pose challenges to collect, store, analyze, and exploit this data in business, including simulations, virtual reality, digital twins, control, and so on [3]. Human factors, such as fatigue indicators, have significant effect on product quality and factory productivity in manufacturing [4]. It is clear that increasing the integration of data (including wearable information) can improve understanding of the production performance. The data silos increase as more and more sensors become available for IIoT devices. The machines are from different vendors, and control mechanisms vary drastically from one vendor to the other, and hence automatic control mechanisms using actuators need to be customized in each case which is a major challenge in industry automation. The data collected by these sensors are owned by either manufacturers or people themselves, and hence, one needs to address the challenges of data privacy on the data captured.

3.4.2 Distributed Ledger Technologies (DLTs)

Distributed ledger technologies (DLTs), such as blockchain, provide a convenient replacement for the central administrator in guaranteeing the integrity of a database. Still blockchain protocols are being defined, and adopting them for mechanization is a long way to go. The lack of interconnected data and common meaning is another limitation for most DLT-based IIoT data platforms.

3.4.3 User Interface (UI) and Machine Learning (ML)

User interface (UI) in IIoT systems is very essential for deriving full advantage of factory automation. UI for IIoT systems, unlike consumer systems, shows complete manufacturing plant's processing workflow, geographic location, and process status in real time integrated with the relevant historical data and decision support system for user interaction. Some examples of smart UI for IIoT applications are shown in Fig. 3.3. Analytic solutions based on machine learning (ML) often operate in real time, adding a new dimension to the industry automation.

Fig. 3.3 Example UIs for IIoT applications (Courtesy: Art Energy)

AI- and ML-based systems are "trained" to use specialized algorithms to study, learn, and make predictions and recommendations from huge data. Predictive models exposed to new data can adapt without human intervention, learning from previous iterations to produce reliable decisions and results. It is believed that with objective way and huge data analytics of machine learning (ML), the solution may exceed human capabilities. Humans recognize sound, objects, and machine surroundings like smoke, color, etc., many times not knowing how. But machines learn to recognize and identify such specific external variables by their characteristics such as pitch, volume, harmonic overtones, etc. This can enhance the predictive analysis of manufacturing processes for maintenance planning and inventory logistics and quality.

The competitive advantage comes from developing machines that do not rely on human sensing, description, intervention, or interaction to solve a new class of problems. This capability opens up new opportunity in many fields, including medicine (cancer screening), manufacturing (defect assessment), and transportation (using sound as an additional cue for driving safety).

3.4.4   Artificial Intelligence

3.4.4.1   High-Performance Computing

IIoT systems require high horsepower computing systems with combination of fixed and programmable accelerators, flexible memories, and networking capabilities suitable for real-time and machine learning algorithms. For example, Intel® Xeon® processors provide a scalable baseline, and the Intel® Xeon Phi™ processor is specifically designed for the highly parallel workloads typical of machine learning, as well as machine learning's memory and fabric (networking) needs. Figure 3.4 shows microarchitecture of Intel's Xeon family processors suitable for IIoT applications.

3.4.4.2   Embedded Systems

The system software for data servers and applications on various platforms along with machine access software applications are developed on the embedded development environments. The embedded development platforms are made developer friendly by many hardware providers with graphical user interfaces (GUIs) with

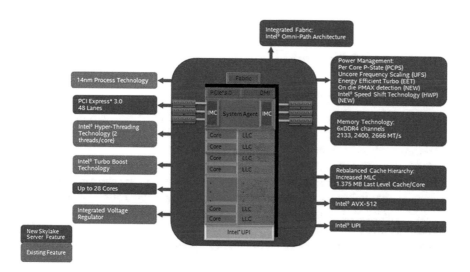

Fig. 3.4 Intel's Xeon family's microarchitecture (Courtesy: Intel)

ready-to-use library of functional blocks and interface functions. There are advanced industrial-grade GUI solutions designed specifically for deeply embedded, real-time, and IoT application development. Full-featured (what you see is what you get) WYSIWYG desktop design tool allows developers to design their GUI on the desktop and generate embedded GUI code that can then be exported to the target hardware platforms. GUI library along with basic functional models will have many device references like industrial control reference, chemical process reference, and automotive reference models.

3.4.4.3 Application Software Development

The application-specific software for IIoT-based automation need to be developed which are user friendly. This involves development of embedded software controlling devices and HMIs that are often a critical part of IoT-enabled products. Depending on the domain for which the IIoT system is developed, the developed code has to be compliant to various standards. For example, embedded code for avionic applications should comply to DO-178B level A safety standard, ISO 26262 safety standard for automotive applications, and functional safety for EC 61508 2010 at T3/safety integrity level (SIL) 3 standards. It is to be remembered that the embedded software development is done for a target hardware platform and then the executable is downloaded onto embedded memory. Hardware platform providers also provide embedded software development platforms wherein software functions including memory management, protection, task management, interrupt and exception handling, multi-processor support, thermal and power management features, debugging, performance monitoring, system management mode, virtual machine extension (VMX) instructions, and data visualization methods.

3.5 IIoT Development Platforms

An IIoT platform should provide:

- Device management software that connects thousands to hundreds of thousands of sensors, industrial machines, and digital systems. IIoT solutions are usually designed to identify failures and facilitate recovery.
- Integration through software development kits, development tools, and APIs to support business processes and enterprise systems across the business. There are significant challenges, given the array of back office applications such as ERP, application performance management, enterprise asset management, computerized maintenance management systems, and more.
- Data management to control and monitor ingestion, storage, accessibility, flow, and security.

- Analytics of data from connected devices, enterprise, and third parties to reveal patterns and optimization of resources.
- The enablement and management of applications to simplify configuring and operating connected assets and that enable *digital twins*.
- Software to allow security audits and ensure compliance, including measures to prevent data loss and detect and act on breaches.
- Support *protocols* relevant to the industrial domain, such as OPC (Open Platform Communications) Unified Architecture.
- Engineering-level robustness to prevent downtime.
- Flexibility with *no-code interfaces*, for example, to allow a range of users to access job-specific applications.
- A combination of cloud computing, on-premise deployment, and *edge computing*.

3.5.1   Commercial IoT Platforms

Well-known commercial IIoT platforms are listed below:

- ATOS IoT
  ATOS IoT is a platform which has to be used for implementing IIoT solutions and offers services for automating manufacturing in packaging, automotive, and process industries. Their offerings include high-end enterprise servers with high-end storage, GPUs with scalable architecture, and cloud solutions. The platform enables connecting data collecting nodes, machines, people, and processes with an agile approach.

- Amazon AWS IIoT Core
  AWS IoT Core is a platform to connect devices to the cloud servers and interact with the other devices and cloud applications. It is a managed cloud service. It provides support for HTTP, lightweight communication protocol, and MQTT protocols. It can process a large number of messages, track, and communicate even when they are not connected on the Internet. It integrates services like AWS Lambda, Amazon Kinesis, and Amazon QuickSight. It allows secure access to the connected IIoT devices.

- IBM Watson IoT platform
  This platform helps capture and investigate the data from IIoT devices, machines, and equipment and has decision support systems for informed decisions for the operators and the management. This platform will optimize operations and resources by providing the correct business insights and bi-directional communication facility. Major features include flexibility, security, capturing real-time data, and providing AI and analytics services.

- Microsoft Azure
  This IoT solution is designed for automation of industries of domains like manufacturing, transportation, and retail to name a few. It provides solutions for

remote process monitoring, predictive maintenance, smart spaces, and connected products. It provides an open platform to build a robust application. There are two solutions for beginners: an IoT SaaS and open-source IoT templates.

- IRI Voracity
  It is a cloud IIoT platform that connects to and integrates sensors, log, and data sources from many other sources, with consolidated (same I/O) data filtering, transformation, cleansing, masking, and reporting. It runs on a wide range of Linux, Unix, and Windows platforms, Raspberry Pi, and Linux mainframe. It transports, replicates, subsets, and leverages IoT data for archival and deriving analytics. It supports for fit-for-purpose data wrangling node to aggregate and anonymize IoT data and feed IoT mining and machine learning nodes. It provides application development and add-on and direct indexing of data for cloud analytics and acts upon IoT data.

- Google Cloud IIoT platform
  Google Cloud provides a multi-layered secure infrastructure for IIoT applications. It helps in improving operational efficiency. It provides predictive maintenance for equipment, solutions for smart cities and buildings, and real-time asset tracking. The solution on this platform can be customized to any particular application.

- Oracle IoT cloud
  With the help of Oracle IoT cloud, one can connect IIoT devices to the cloud, perform analysis of data in real time, and perform integration of data with enterprise applications or web services. It supports integration with Oracle and non-Oracle applications and IoT devices using application programming interfaces (APIs). Platform allows creating an IoT application to connect a device using JavaScript, Android, iOS, Java, and C POSIX. Many logistical supply chain, ERP, HR, and customer experience applications can be built on this platform. It provides features like device virtualization, high-speed messaging, and endpoint management to connect.

- PTC: ThingWorx
  ThingWorx helps in managing the development life cycle for IoT applications. It provides flexibility to access data and IoT from on-premises, off-premises, and hybrid environment. Use of ThingWorx platform increases uptime, reduced costs, provides role-based visibility and control, and improved compliance to process flows. Main features are adding devices in the network, analyzing captured and collarated data, and data visualization. Industrial IoT and application data is accessible from on-premises web servers and off-premises cloud applications and as hybrid environments.

- MathWorks ThingSpeak
  ThingSpeak is an IoT analytics platform service from MathWorks that allows to aggregate, visualize, and analyze live data streams in the cloud. One can send

data to ThingSpeak™ from the IoT or IIoT devices, create instant visualizations of live data, and send alerts using web services like Twitter®.

- Cisco IoT cloud
  Cisco IoT cloud connect is a mobility cloud-based software suite. This IoT solution is for mobile operators. It will fully optimize and utilize the network. Cisco provides IoT solutions for networking, security, and data management with granular and real-time visibility, provides updates for every level of the network and IoT security by protecting the control system from human errors and attacks, and increases visibility and control by defending malware and intrusion and centralized security controls.

- Altair SmartWorks
  Altair SmartWorks provides an end-to-end IIoT development platform. It provides a platform as a service. It will help you to connect devices, collect data, manage devices and data, and build and run the applications. It provides functionalities like device management, listeners, rules, custom alarms, triggers, and data export. It has an open architecture.

- Siemens: MindSphere
  MindSphere is industrial IIoT as service solution. Using advanced analytics and AI, MindSphere powers IoT solutions from the edge to the cloud with data from connected products, plants, and systems to optimize operations, create better-quality products, and deploy new business models.

3.6 IIoT Subsystems

The major subsystems of the IIoT solutions as evident from architecture and platforms are device subsystem, high-performance computing subsystem, communication systems, and cloud subsystems backed with right size memory at each stage of the subsystems.

3.7 Communication Technologies

Communication technologies suitable for IIoT solution are fully digital broadband technologies. The communication technologies for IIoT solutions mean to connect devices with all types of information including voice, video, data, connecting drones, robots, or sensors over secure, reliable, and easy-to-use means. Additionally, it has to be secure and resilient. These will be dealt in detail in later chapters.

References

1. Scheuermann, C.; Heinz, F.; Bruegge, B.; Verclas, S. Real-Time Support During a Logistic Process Using Smart Gloves. In Proceedings of the Smart SysTech 2017; European Conference on Smart Objects, Systems and Technologies, Munich, Germany, 20–21 June 2017; pp. 1–8.
2. Hao, Y.; Helo, P. The role of wearable devices in meeting the needs of cloud manufacturing: A case study. Robot. Cim.-Int. Manuf. 2017, 45, 168–179.
3. Vrchota, J.; Pech, M. Readiness of Enterprises in Czech Republic to Implement Industry 4.0: Index of Industry 4.0. Appl. Sci. 2019, 9, 5405.
4. Neumann, W.; Kolus, A.; Wells, R. Human Factors in Production System Design and Quality Performance - A Systematic Review. IFAC-PapersOnLine 2016, 49, 1721–1724.

Chapter 4
IoT Product Design: A Case Study

Introduction to Design Case

Environment monitoring is necessary for a number of reasons major among them is sustainable growth of the humanity. In general, environment monitoring gives us insight into the present levels of harmful or potentially harmful pollutants discharged to the environment by various sources and helps find the effect of these pollutants on the human health and climatic changes. Recent advancement in IoTs and wireless sensor networks coupled with artificial intelligence play a major role in making environment monitoring a smart monitoring. Wireless sensor networks (WSN) use IoT devices networked using wireless technologies like Bluetooth and WLAN. WSN can monitor the environment by monitoring temperature, humidity, noise and pollution level and controlling them. Smart environment monitoring deploys multi-model spectral data captured through remote sensing satellite imaging and IoT-based environment monitoring systems. Similar concepts can be applied to green house farming to control environment inside the greenhouse and control them as required by the crop and increase the produce. Figure 4.1 shows a cloud-based smart environment monitoring system.

4.1 Design Case Study: Smart Environment Monitoring

Monitoring temperature and humidity is one of the important components of a smart environment monitoring system. A reference design, an IoT-based smart environment monitoring system, to monitor temperature and humidity, is discussed. The proposed IoT system is required to capture and update the environment parameter database on the cloud server with the measured parameters on need basis and periodically at preset times so that user can access the data from anywhere anytime.

Fig. 4.1 Smart
environment monitoring
system using cloud-
connected IoT systems

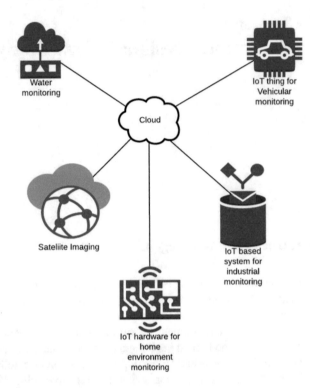

4.2   A Case Study: Product Requirements

The requirements of smart IoT system for environment monitoring are the following:

1. Sense temperature and humidity of the surrounding area on demand or at regular interval.
2. Remotely tune captured parameters by controlling heating or cooling devices in a closed environment or to get trends and statistics.
3. The sensed data is periodically uploaded to cloud storage.
4. Remote access to the data from cloud server when needed from anywhere.

4.3   Proof of Concept

The proof-of-concept IoT system consists of:

- Sensors for measuring temperature and humidity
- Processor to process the data
- Communication interface to transfer the processed data onto the cloud platform
- Data visualization setup for the data stored in the cloud
- Mobile/system application to retrieve the data from anywhere, anytime

The architecture of the POC IoT-based smart environment monitoring system is shown in Fig. 4.2. It consists of an IoT device whose internal block diagram shown in light blue has microcontroller (uC) on-chip module and Wi-Fi module with humidity and temperature sensor connected. The system is interfaced with a Wi-Fi module which enables two-way wireless data communication to write the data into the cloud storage, in this case, MathWork's ThingSpeak cloud.

The above system can be built by using the following modules shown in Fig. 4.3.

- DHT11 temperature and humidity sensor
- ESP8266 Wi-Fi module
- Arduino Uno board
- Power supply module

The IoT device hardware in this design will have embedded firmware stored in the IoT device for each of the modules except the power supply.

The stored data on cloud can be accessed by any Internet-capable device like a laptop computer or a smartphone or an iPad from anywhere which support Message Queuing Telemetry Transport (MQTT) protocol API or representational state transfer (REST) APIs as in Fig. 4.2. MQTT and REST protocols are M2M communication protocol on TCP/IP layer of Internet access. Cloud application also performs data processing and data visualization functions in a user-friendly way.

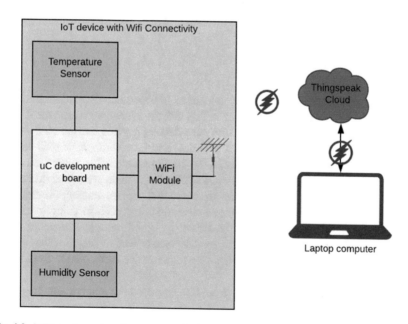

Fig. 4.2 IoT-based smart environment monitoring system

DHT11 temperayure &
Humidity Sensor

EPS 8266 Wifi module

12V to 3.3V and 5V power
supply module

Arduino Uno development
board

Fig. 4.3 Hardware components used in the IoT device for smart environment monitoring

4.3.1 IoT Device

The experimental setup to prove the concept of IoT-based environment monitoring system is shown in Fig. 4.4. As shown in Fig. 4.4, the system with integrated development environment (IDE), where firmware development is carried out, is interfaced with the Arduino board using serial interface through UART signals. Temperature-humidity sensor DH11 and the Wi-Fi module EPS8622 are interfaced to Arduino board through GPIO signal as shown in the interface signal description Table 4.1. The firmware program when executed will capture temperature-humidity data from the DH11 module every one minute and transmits it to ThingSpeak platform on the cloud server through the EPS8622 Wi-Fi module connected to the Arduino Uno board. The ThingSpeak cloud service is used to visualize the updated data transmitted from anywhere anytime on smart device connected to the Internet by the registered user.

The input-output interface signals from different modules and systems are shown in Table 4.1.

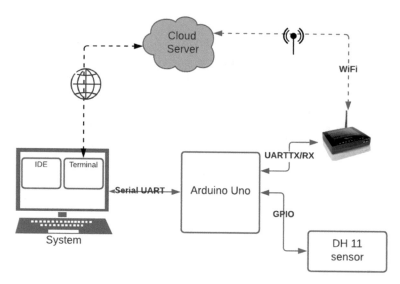

Fig. 4.4 Environment monitoring system setup

Table 4.1 Interface signal description

Sl. no.	Signal name	Signal description	Remarks
DH11			
1	Vcc	5 V power supply to sensor module	Power supply
2	Out	Sensed temperature or humidity output	Sensor data output
3	Gnd	0 V	Ground
WI-FI			
1	Gnd	0 V	Ground
2	Txd	UART TX	Data transmitted to Wi-Fi module
3	GPIO 2	NC	Not used in this design and hence not connected
4	CHIP_ PD	5 V	Chip enable High: Normal Low: Low power mode
5	GPIO 0	NC	High: Normal functional mode Low: sets Wi-Fi module to boot loader mode
6	Ext_Rstb	NC	Active low reset
7	Rxd	UART Rx	Data received by Wi-Fi module
8	Vcc	3.0 to3.6 V	5 V external power supply connected to 3.3 V on board regulator which converts it to 3.3 V
Arduino Uno			
0	GPIO	UART_RX	From software terminal application to get data. Serial in signal from UART
1	GPIO	Hardware UART_TX	To software terminal application to display the sensed output. Serial out signal through UART
10	GPIO	Software UART RX	Data received from Wi-Fi module
11	GPIO	Software UART TX	Data to be transmitted to Wi-Fi
2	GPIO	Digital input	Sensed data from DHT11 sensor

4.3.2   IoT Device Firmware

The IoT firmware development is done using Arduino IDE. The header files corresponding to the hardware modules interfaced to the Arduino board are called shields. These shields are included in the firmware program in the Arduino IDE. This reference design needs Wi-Fi shield, DH11 sensor shields which are included in the firmware program with the *#include* commands, and application-specific code for connecting to Wi-Fi modem and accessing cloud services.

4.3.2.1   Firmware Code

The embedded application-specific firmware for capturing the temperature and humidity is given in Fig. 4.5:

Firmware program starts with including necessary libraries using *#include* command, followed by connectivity information (which interface pin is connected to which pin of external module) and then creating the object corresponding to the connected external module as done in the first part of the code shown.

```
  include <SoftwareSerial.h>.
//include serial interface shield
  #include "DHT.h"
//include DHT.h file

  #define dht11Pin              2
// Digital pin connected to the DHT sensor
  #define DHTTYPE         DHT11
// DHT 11

  DHT dht (dht11Pin, DHTTYPE);
//Create DHT object
  SoftwareSerial esp (10, 11);
// RX, TX
```

The second part of the program is the **setup** where the *initialization* of the module is done. The data rate of serial interface, assigning default logic states to interface signals are done in Setup part of the program which is shown in the code below:

```
include <SoftwareSerial.h>.              //include serial interface shield
#include "DHT.h"                         //include DHT.h file

#define dht11Pin        2                // Digital pin connected to the DHT sensor
#define DHTTYPE      DHT11               // DHT 11

DHT dht (dht11Pin, DHTTYPE);            //Create DHT object
SoftwareSerial esp (10, 11);            // RX, TX

void setup ()
{
  String network =  "wifi_network";     // write the name of our network
  String password = "our_password";     // write the password of our network here.
  dht.begin();
  Serial.begin(115200);                 // We are starting our communication with the
serial port.
  Serial.println ("Started");
  esp.begin(115200);                    // We are starting serial communication with
ESP8266.
  esp.println("AT");                    // We do the module control with the AT command.

  while (!esp.find("OK"))
  {
    Serial.println("Resend");
    esp.println("AT");
  }
  Serial.println("OK Command Received");
  esp.println("AT+CWMODE=1");           // We set the ESP8266 module as a client.
  while (!esp.find("OK"))               // wait until we recieve response from Wifi
module
  {
    Serial.println("Setting is ....");
    esp.println("AT+CWMODE=1");
  }
  Serial.println("Set as client");

  Serial.println("Connecting to the Network ...");
  esp.println("AT+CWJAP=\"" + network + "\",\"" + password + "\""); // We are connecting to
our network.
  while (!esp.find("OK"));                             // We wait until it is connected to the
network.
  Serial.println("connected to the network.");
  delay(1000);
}
void loop()
```

Fig. 4.5 IoT firmware program for monitoring environment (temperature-humidity) parameters

```
{
 String ip = "184.106.153.149";                              //Thingspeak ip adress
 float temperature, humidity;

 delay(2000);
 esp.println("AT+CIPSTART=\"TCP\",\"" + ip + "\",80");      // We connect to Thingspeak.
 if (esp.find("Error"))
 { // We check the connection error.
   Serial.println("AT+CIPSTART Error");
 }

 humidity = (float)dht.readHumidity();
 temperature = (float)dht.readTemperature();

 // Serial.print(F(" Humidity: "));
 // Serial.print(humidity);
 // Serial.print(F("% Temperature: "));
 //Serial.print(temperature);
 //Serial.print(F("C "));


 String veri = "GET https://api.thingspeak.com/update?api_key=CRF9H8AJ0DC5UFL7";
                                    // Thingspeak command. We write our API key
                                    // in the keypart.
 veri += "&field1=";
 veri += String(temperature);              // The temperature variable we will send
 veri += "&field2=";
 veri += String(humidity);                 // The moisture variable we will send
 veri += "\r\n\r\n";

 esp.print("AT+CIPSEND=");                  // We give the length of data that we will send to ESP.
 esp.println(veri.length() + 2);
 delay (2000);

 if (esp.find(">"))
 {                                          // The commands in it are running when ESP8266 is
ready..
   esp.print(veri);                         // We send the data.
   Serial.println("Data sent.");
   Delay (1000);
 }
 Serial.println("Connection Closed.");
 esp.println("AT+CIPCLOSE");                // We close the link
 delay (1000);                              // We wait 1 minute for sending new data.


}
```

Fig. 4.5 (continued)

```
void setup ()
{
   String network =    "smile_to_connect";
// We write the name of our network  here.
   String password = "our_passward";
// We write the password of our network here.
   Serial.begin (115200);
// We are starting our communication with the serial port.
}
```

The third part is the "*loop function*" which is executed forever.

```
void loop()
{
  float temperature, humidity.
  humidity = (float) dht.readHumidity();
  temperature = (float) dht.readTemperature();
}
```

The data transmitted from the Arduino IDE setup is also displayed on the terminal connected through COM port to the Arduino board UART module. The data display on the COM port terminal is shown in Fig. 4.6.

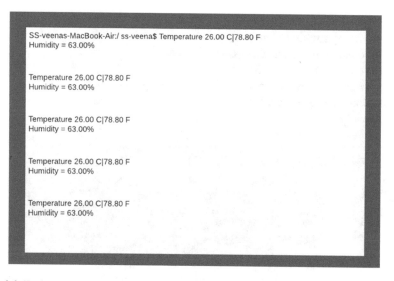

Fig. 4.6 Environmental data captured by the IoT-based system displayed on the COM port terminal

4.3.2.2   Prototype Design

Prototyping a product is a first step of product design. Product prototyping is product design in small numbers to confirm the manufacturability of the product. Product design is the most important aspect of IoT device, seen as the physical makeup of the hardware. IoT devices have to be safe, convenient to mount and install, and aesthetically appealing in most of the applications. Key factors considered for the product design are best user experience, ease of use, ease of adoption to everyday life, consistent user experience in traversing digitally from system to cloud applications, and personalized user experience. Product design should use materials whose properties and behavior when exposed to environmental conditions are well known. Features such as waterproofing and biocompatibility of the product depend on the material properties.

Product design and electronic design should go hand in hand. The product design flow is shown in Fig. 4.7.

Product design is carried out using CAD software tools, which have the following features:

- Assess manufacturability and product assembly including mechanical parts.
- Create precise geometry of parts and model.
- Support assembly modelling.
- Generate 2D and 3D graphics of parts as per international standards ASME, ISO, and JIS.
- Automatically generate drawings of parts with assembly notes.
- Automatically generate bill of materials with assembly notes.
- Software tool performance: fast load of complex parts with lightweight representation.
- Analysis module for thickness.
- Mold filling simulation module.
- Auto creation of holes, springs, and fasteners.
- Auto optimization capability.
- Create parts by reverse engineering of existing product capture.
- Top-down design or teardown design viewing and drawings.

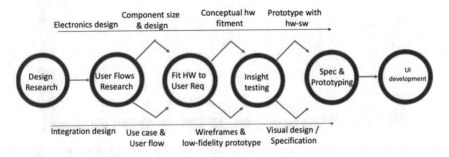

Fig. 4.7 Product design flow

Fig. 4.8 NetBotz Wireless Temperature and Humidity Sensor product (Courtesy: Schneider)

Many product design tools are available. Creo, Inventor, Fusion 360, Illustrator, and Onshape are some of the popular product and machine design tools.

Figure 4.8 shows the commercially available NetBotz Wireless Temperature and Humidity Sensor-based IoT, from Schneider [3]

4.3.2.3   Cloud Storage and Analytics Setup

Most of the cloud platforms discussed in Chapter 2 provide free access to the cloud storage and analytics with simple user registration. This enable the IoT user to access various services like access to related data hosted by others at a price. For example, the owner of smart environment monitoring system can access from the cloud service provider from his subscription and he can also subscribe to services provided by others for related geographic data like coordinates and terrains. These services are provided in "software as a service (SaaS)" model in a way that makes environment information more meaningful.

4.3.2.4   Registration to ThingSpeak Cloud

Access to cloud services is by user registration using user details and with valid email address. The cloud service providers like ThingSpeak, Amazon, IBM, and Cisco send the verification link to check the validity of the email which has to be used to login into the cloud for confirmation.

IoT case study uses ThingSpeak cloud platform from MathWorks [4] for remote access and storage. After registration to cloud service, to collect, store, analyze, and display the data, one has to generate a channel. ThingSpeak stores device data in channels as messages. User data from IoT devices are collected, analyzed, and controlled using these channels in ThingSpeak cloud service. Normally, one connected device will require one channel on ThingSpeak. Any other device on the Internet which supports these protocols can access the stored environment information via the channel.

4.3.2.5   ThingSpeak Cloud Server Access

Access to cloud-based ThingSpeak channel is through application-specific interfaces (APIs). APIs access web server through the Internet and can use HTTP, REST APIs, and MQTT protocols. These protocols are discussed in detail in M2M communication part of the book. To read and write to a ThingSpeak cloud channel, the application sends requests to the ThingSpeak server by issuing HTTP commands, publishing MQTT messages. Each ThingSpeak channel can have up to 8 fields of 255 characters of data, in either numeric or alphanumeric format. A channel also has location information and a status update field. Each channel data entry is stored as a message of eight fields of data, with a date and timestamp. One can retrieve stored data by time or by entry ID.

The IoT device data gets updated to the user channel through the following highlighted code in the device firmware. This is REST API for updating the channel with new measured data. The highlighted text in the statement is the API key for writing into the channel.

```
{
GET https://api.thingspeak.com/update?api_key=CJMNXT0RFLNKH7J1&fie
ld1=&field1&field2=&field2
}
```

Data retrieval to IoT device is done using code embedded in the firmware below:

```
{
PUT https://api.thingspeak.com/update?api_key=CJMNXT0RFLNKH7J1&fie
ld1=&field1&field2=&field2
}
```

Data set can be viewed by any other authorized client anytime and from anywhere by logging into the ThingSpeak cloud with proper credentials. The snapshot of the channel data is shown in Fig. 4.9.

The cloud service providers offer many services to export the data set in many formats like the tabular form in CSV, JSON, and XML formats for further viewing and processing from anywhere.

Table 4.2 shows the temperature and humidity of the environment in CSV file format which is retrieved from the cloud service provided for the registered user.

Channel Stats

Created: about a month ago
Last entry: about 18 hours ago
Entries: 18

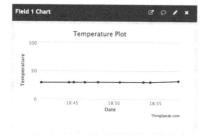

Fig. 4.9 Snapshot of the environmental data: temperature and humidity on the cloud server. (Courtesy: MathWorks ThingSpeak output)

Table 4.2 Environmental data set as updated periodically in the ThingSpeak cloud server exported in CSV format

Feed			
created_at	entry_id	Field1	Fleld2
2021-01-26 12:18:32 -0500	1	25	
2021-01-26 12:20:46 -0500	2	27	
2021-01-26 12:21:29 -0500	3	23	
2021-01-26 12:22:14 -0500	4	23	89
2021-01-26 12:22:38 -0500	5	22	95
2021-01-26 12:23:15 -0500	6	26	96
2021-01-26 12:25:11 -0500	7	24	96
2021-01-26 12:35:48 -0500	8	21	100
2021-01-27 11:32:37 -0500	9	26	95
2021-03-05 08:11:19 -0500	10	28.80	21.00
2021-03-05 08:14:36 -0500	11	29.10	20.00
2021-03-05 08:15:12 -0500	12	29.20	20.00
2021-03-05 08:16:32 -0500	13	29.20	19.00
2021-03-05 08:18:06 -0500	14	29.30	19.00
2021-03-05 08:20:46 -0500	15	29.40	19.00
2021-03-05 08:23:33 -0500	16	29.40	19.00
2021-03-05 08:24:24 -0500	17	29.40	20.00
2021-03-05 08:27:47 -0500	18	31.40	49.00

4.3.2.6   Product Design

Once the prototype is validated and accepted by engineering, marketing, and manufacturing, the product is ready for manufacturing in large numbers. It also means
that the product is ready to be launched in the market. This stage is considered as
engineering success. From this phase onward, it will be production management
and marketing activity. The marketing and sales team will get product orders, and
the production team will work and execute the orders. Market success depends on
product quality, market acceptance, customer feedback, marketing campaigns, and
promotions.

4.3.2.7   Product Validation and Testing

Prototype and product validation procedures are identical except that the large-scale
product testing is carried out by production teams, whereas for prototypes, it is done
by engineering teams. IoT devices and systems are very complex involving many
technologies. So, it is essential to test the system at all stages. The complete IoT
ecosystem solution for ambient environment monitoring of temperature and humidity is discussed here can be represented as in Fig. 4.10.

Product validation during development happens at all stages involving hardware,
firmware and, when integrated with the applications on the Internet, integration testing is carried out. Once the system passes preliminary test at this level, it is necessary to look at the overall test framework. The IoT system tests at the product level
are classified as follows:

- Functionality test
- Compatibility test
- Stress and scalability test
- Data integrity test

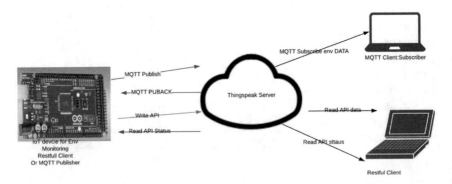

Fig. 4.10 Environment monitoring system for monitoring ambient temperature and humidity

- Security test
- Performance test

All the tests mentioned above are explained under the context of the reference design of environment monitoring device.

Functionality Test

The IoT device in reference design is tested for the supported functionalities implemented.

Temperature Measurement: The device should be able to measure the temperature to the accuracy claimed under different conditions. This test requires golden reference temperature standard to declare it to be functioning correctly.

Initially, measurements are taken for major use case conditions and calibrated against the standard values. Once the device is calibrated, the trend in deviation is observed closely to see whether the device is drifting in any particular direction or on both sides with respect to the standard. These data from the device under test and standard are recorded for a large number of cases, and the data is statistically analyzed by calculating mean difference and standard deviation. Formal analysis is carried out by plotting the Bland-Altman chart to check if there is any skewness in any specified range of temperature. Skewness is corrected by selective calibration. After product calibration, the design is ready for integrated test. Similarly, functional test and analysis are carried out for humidity also.

Compatibility Test

IoT devices get connected to cloud server through a protocol and to other devices on the IoT cloud platform. As a result, checking the data compatibility of IoT system with cloud is important. Major tests include data write to cloud, data retrieval from the cloud, and connection to mobile smartphone application (if claimed). Data integrity could be affected due to a trivial facts like mismatch in endianness. That should be tested and confirmed. Also, when the cloud-based web server data is retrieved onto smart mobile applications, data visualization could be a problem as the screen size gets shrunk to smart mobile display from large web display. This has to be tested for data visualization integrity.

Stress and Scalability Test

Reliability and scalability are important for building an IoT test environment which involves simulation of sensors by utilizing virtualization tools and technologies. This is carried out repeating the tests for large number of times for different scenarios. This is called stress testing.

Data Integrity Test

It is important to check the data integrity in IoT device as it handles a large amount of data. Since IoT system includes a third-party service like the cloud, it is essential to get the error-free information and pass it on to the end user who is trying to access the data.

Security Test

In the IoT environment, many users are accessing a massive amount of data from the same cloud. Thus, it is important to validate user authentication and data privacy controls as part of security testing.

Performance Test

Performance tests include speed tests, back-to-back data test, and memory round off tests, which in turn affect the data integrity also. The entire system has to be tested for these scenarios at all interfaces. Performance testing is also important to create strategic approach for developing and implementing an IoT testing plan.

Table 4.3 shows the relevance of the tests for different components of an IoT system.

4.3.2.8   Safety and Environment Tests

Since the IoT devices are used as consumer and industry applications, they have to comply with certain safety and environment safety standards defined by the regulatory authorities, for example, RF radiation tests, leakage current tests, and battery tests. There are exclusive test houses, which have test facilities such as special chambers in which device under test (DUT) is kept and powered externally. The test scenario is created, the response observed and validated. These test houses certifies the products for compliance to identified relevant standards. TÜV Rheinland is one

Table 4.3 Relevance of test types at different components of IoT system

IoT elements Testing type	Sensor	Application	Network	ThingSpeak cloud
Functional testing	True	True	False	False
Usability testing	True	True	False	False
Security testing	True	True	True	True
Performance testing	False	True	True	True
Compatibility testing	True	True	False	False
Services testing	False	True	True	True
Operational testing	True	True	False	False

such global test house. There are many state affiliated accreditation boards for testing and calibration. These test houses in India offer test facilities and certify the devices for safety and environmental conditions as per the standards in each application domain.

IoT Testing on the Internet

Similar to Google and Bing, there are search engines on the Internet to track the IoT connectivity such as *Shodan* [5] (www.shodan.io) and *Thingful* [6] (www.thingful. net). They are different from Google and Bing in that they search devices on the Internet unlike Google and Bing, which search the World Wide Web. They provide insight into facts like number of IoT devices on internet, country having most IoT systems etc. This information can be used for providing customized security to the user device data on the Internet. Thingful, a search engine for the Internet of Things, allows secure interoperability between millions of objects on Internet.

All the test conditions relevant for the reference design are listed in Table 4.4.

4.4   IoT Device Using VLSI System on Chip (SoC)

IoT device of the smart environment monitoring system can be realized using system on chip instead of integrating multiple discrete modules on PCB. This means that the Arduino development board modules [2], Wi-Fi module, and sensor modules can be integrated on to single chip using VLSI fabrication technology. The advantage of doing this is miniaturization, reduced power consumption, reduced complexity, higher reliability, and lower cost per device when manufactured in large numbers. In IoTs and IIoTs where devices are to be mounted at remote locations, maintaining them remotely is very challenging and expensive and all the more challenging when the device is battery operated because of low processing power and storage in the device. SoCs are effective and cheaper alternatives as they provide more storage and processing powers due to miniaturization.

Integrating different functions include integrating digital signal processing and data processing blocks, analog blocks, mixed signal functions like ADCs and DACs, and wireless processing blocks in one IC chip. Examples of this kind of SoCs are EFR32MG21 Mighty Gecko Multiprotocol Wireless SoC [7] from Silicon Labs. The internal block diagram of the SoC is shown in Fig. 4.11.

There are IIoT SoCs available with integrated functions from many vendors for human-machine interface and multiple wireless functions for industrial automations. BGX13P22GA [8] is one such example for Industrial IoT. Figure 4.12 shows the block diagram of BGX13P22GA from Silicon Labs.

Table 4.4 Test condition relevant for IoT for smart environment monitoring system reference design

Test categories	Sample test conditions
Component validation	• Device hardware sensor for temperature reading and humidity reading • Communicating the read data from sensor onto ThingSpeak cloud through Wi-Fi module • Embedded software for cloud access and monitoring the temperature and humidity data • Cloud infrastructure for registration, channel creation, and data visualization • Network connectivity through Wi-Fi module • Third-party software like web application through APIs • Sensor testing for sensed data for accuracy and reliability • Command testing for MQTT or REST protocol • Data format testing for data integrity during communication from IoT device to ThingSpeak cloud and from cloud to any other device • Robustness testing when there is sudden variation in temperature and humidity during rainfall, etc. • Safety testing as per the standard requirement
Function validation	• Basic device testing for proper module functions and interconnections • Testing between IoT devices (end-to-end tests) • Error handling (negative tests, e.g., when the temperature is out of range or cloud connection not established). Negative testing can be even when the IoT device detects temperature changes faster than cloud access. There is possibility of data loss • Valid calculation for data visualization and data endianness
Conditioning validation	• Manual conditioning different modes like temperature testing and humidity testing modes • Automated conditioning when data is written at the frequency cloud data is updated • Conditioning profiles data visualization
Performance validation	• Data transmit frequency: continuous data read/writes from device to cloud • Multiple request handing stressing the protocol tests • Synchronization by having temporary data buffer if device is faster (which is usually the case) than the cloud update • Interrupt testing when any critical event of cloud server breakdown or device fault is detected • Device performance • Consistency validation
Security and data validation	• Validate data packets for registered and unregistered devices • Verify data loses or corrupt packets due to frequency mismatches • Data encryption/decryption access denials for unauthorized devices • Data values for integrity at all stages • Users' roles and responsibility and its usage pattern • User profiling for optimization
Gateway validation	• Cloud interface testing for Wi-Fi communication • Device to cloud protocol testing Wi-Fi communication • Latency testing for performance evaluation for any data loss
Analytics validation	• Sensor data analytics checking data visualization • IoT system operational analytics • System filter analytics selective update of the parameters • Rules verification
Communication validation	• Interoperability data retrieval from any third-party device on the Internet • M2M or device-to-device end-to-end testing • Protocol for Wi-Fi and MQTT or HTTP or REST

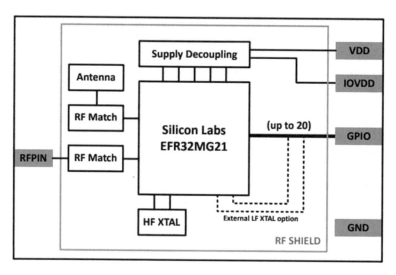

Fig. 4.11 Block diagram of MGM210P Mighty Gecko SoC. (Courtesy: Silicon Labs)

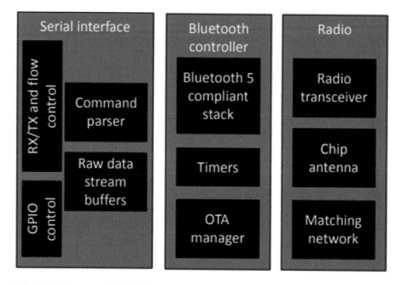

Fig. 4.12 Block diagram of BGX13P22GA SoC for IIoT and many IoT applications. (Courtesy: Silicon Labs) [8]

As it can be seen, in these SoCs, the application processor, wireless communication processors with RF transceiver, antenna, digital baseband, MAC, and power management blocks are integrated on a single chip. It requires sensors, to be connected to serial interface, user interface through GPIOs and battery have to be connected to realize miniaturized IoT device.

There are also SoCs that support multiple wireless protocols like Wi-Fi, Bluetooth, and Zigbee in a single chip, and depending on the application, suitable wireless block can be enabled. A few SoCs also support encryption functionality, which is essential for IoT devices.

It is to be understood that SoC-based IoT implementations are growing in demand. Development of SoC itself is a very specialized, expensive (NRE cost), and time-consuming process.

References

1. https:developer.android.com
2. www.arduino.org
3. www.schneider.com
4. www.mathworks.com
5. www.shoden.io
6. www.thingful.net
7. EFR32MG21 Mighty Gecko Multiprotocol Wireless SoC Family Data Sheet
8. BGX13P Bluetooth ® Xpress Module Data Sheet, Silicon Labs

Chapter 5
Sensors, Actuators, and Hardware Accelerators

5.1 Introduction to Sensors and Actuators

IoT and IIoT devices interact with the environment through sensors and actuators. Sensors convert magnitude of physical parameters to electrical signals, which can be easily processed. For example, a temperature sensor will convert temperature into an electrical parameter like resistance or current in a circuit, which is converted to a voltage. The voltage is measured directly or indirectly to compute the temperature which is displayed or further processed by a computing device after analog-to-digital conversion. These are called transducers. Actuator converts the electrical signal to a control action like turning ON?OFF the switch or relay switch ON/OFF, a relay in control system. An example is automatic control of door opener in the mall or automatic control of air cooling system in the room.

5.1.1 Sensor

Sensor is a device that responds to its environment and whose characteristics vary according to a physical parameter, such as pressure, position, temperature, deformation in material (strain), vibration, angle, acceleration, and composition of a material like blood. Transducers are sensors which produce an electrical output according to the magnitude of a physical parameter. These are often used as synonyms as most sensors are transducers. Sensors are vital for IoT, Industry 4.0 and data-driven technologies. A large number of sensors are used in medical, industrial automation, agriculture, automotive, aerospace, and defense applications. They act as interface between physical entities and electronics in many applications.

© The Author(s), under exclusive license to Springer Nature Switzerland AG 2021
V. S. Chakravarthi, *Internet of Things and M2M Communication Technologies*,
https://doi.org/10.1007/978-3-030-79272-5_5

5.1.2 Classification of Sensors

There are many types of sensors. A few of the sensors are listed below.

5.1.2.1 Flow Sensors

Flow sensors measure flow rate of liquid or gas in cubic meter per second. Gas and liquid flow sensors are used in many chemical and beverage companies where it is essential to control the flow of gas or liquid. Flow sensors and switches are also used in water turbines and electricity generators in IIoT devices. A common type of liquid sensor is a hall sensor, which produces electrical pulses when a wane/wheel rotates at a speed proportional to the flow. A magnet attached to the wheel produces electric voltage pulse as it passes past a hall sensor in which current is flowing. This voltage is measured to derive the flow rate of the liquid. There are many commercial flow sensors integrated with switches which can control water, acid, and other chemicals. Gas sensors often work on the principle of cooling a heated wire; the higher the flow, the larger the cooling. For gas flow measurement, the density of gas is required which has to be separately provided.

5.1.2.2 Force Sensors

Force sensors or load cells measure the force applied on it in Newtons or kilograms. There are many types of force sensors, the most common among them being strain gauge sensor or piezoelectric sensor. In strain gauges, resistance varies with strain/force in a resistance bridge which is used to derive a voltage proportional to force/strain. Piezoelectric sensor produces voltage proportional to force applied on it.

5.1.2.3 Humidity Sensors

Humidity sensors measure relative humidity in percentage which can be converted to absolute humidity content if temperature is known. So, invariably humidity sensor is accompanied by a temperature sensor. Simple resistive or thermal conductivity sensors are used as humidity sensors. The conductivity of non-metallic material changes with the water vapor content of its surroundings. Humidity sensors are used in environment monitoring IoT systems for precision agriculture applications. These sensors are also made up of capacitive technology where a porous dielectric used changes its relative permittivity according to humidity. These are used to detect accurate measurement of dew point.

5.1.2.4 Pressure Sensors

Pressure is force per unit area in newton per square meter. All force sensors are pressure sensors. Pressure sensors are used in IoT and IIoT systems in industry automation, healthcare, and chemical industries. Common types of pressure sensors are MEMs sensors.

5.1.2.5 Photo Optic Sensors

Key parts of non-invasive patient monitoring IoT devices are photo optic sensors. For example SpO_2 measurement in body vital monitoring uses photo optic sensor using pulse oximetry method. They need to be of high precision as they are used in patient monitoring.

5.1.2.6 Water Level Sensors

Water level indicators are used in home automation systems using IoT. They also find application in gas and chemical industry applications. These sensors are resistive or capacitive sensors.

5.1.2.7 Ultrasonic Sensors

Ultrasonic sensors measure distance from the time taken for reflected wave to arrive. This measures liquid level despite variations in transparency, viscosity, color, or dielectric constant. Applications include air bubble detection for medical pumps; point and continuous level sensors for the semiconductor and high purity applications like clean rooms; and point level sensors for a variety of process control applications.

5.1.3 Actuators

Actuators in the IoT devices are electromechanical devices that provide controlled movements or positioning when subjected to electrical stimulus. These could be valves or switches, which further control the flow of a process in industrial automation or simply liquid/gas flow. The physical movement can be linear or rotary, depending on which actuators are classified as linear or rotary. Linear actuators usually have a push and pull function with rigid chains. Rotary actuators provide rotary motion that controls valves such as ball valves or butterfly valves that control flow of liquid in a pipe. Figure 5.1 shows a ball valve and a butterfly value.

Actuators come in different power configurations and sizes depending on the applications. Types of actuators are linear electrical actuators, rotary linear

Fig. 5.1 Ball valve and butterfly valve (Courtesy: Kirloskar Ltd.)

actuators, micro-actuators, silicon micro-grippers, micro-heaters, and micro-motors. Actuators find applications in a variety of boilers and heaters in both industrial and domestic environments, and the micro-mechanical actuators find application in biomedical instrumentations. Micro-tools like micro-grippers are used in minimally invasive diagnostic medical applications during robotic surgery. Advancement in micro-machining processes and suitable materials have given rise to the new domain of engineering called microelectromechanical systems (MEMS) with which the micro-mechanical devices like micro-gears, micro-heaters, and micro-switches are made. The phenomenal potential of this technology is yet to be tapped. The MEMS devices are highly reliable, easy to handle and of miniature size giving rise to smart IoT devices. Few of the commercial MEMS actuator devices are shown in Fig. 5.2.

Figure 5.3 shows ADRF5046 from Analog Devices [1], released in 2019, which is MEMs based a single-pole, four-throw (SP4T) RF switch for controlling signals from 100 MHz to 44 GHz.

5.2 Signal Conditioning of Sensor Output

Signal conditioning of the sensor circuit is the process of converting the output voltage/current from the sensor to a form suitable for further processing by computing machines in data acquisition systems. Methods used to make sensor outputs suitable for processing include amplification, raising signal-to-noise ratio, or matching the impedance to avoid reflection/loading. Typically, these circuits are analog or mixed signal circuits which are designed as a part of data acquisition system. For example, small strain gauge output signal is converted to digital data, which need biasing and

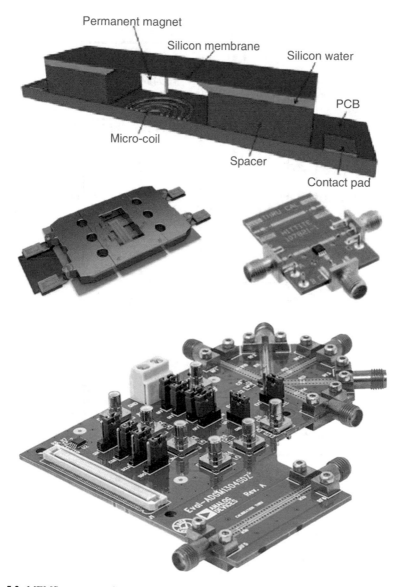

Fig. 5.2 MEMS actuators. (Courtesy: RF Switch from Analog Devices)

amplification. Figure 5.4 shows the signal conditioning circuit in data acquisition block of IoT devices.

A simple data acquisition system is shown in Fig. 5.5. The analog outputs from sensors are multiplexed to a single analog amplifier after converting to digital signal. It is then stored in memory for further use. This architecture saves on the processor hardware.

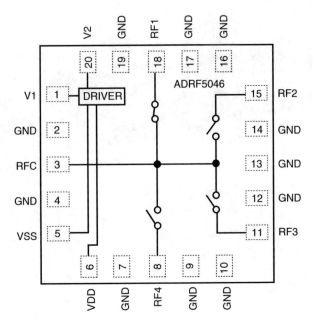

Fig. 5.3 ADRF5046 RF switch from Analog Devices (Courtesy: Analog Devices)

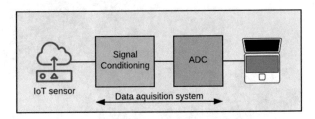

Fig. 5.4 Signal conditioning in IoT device

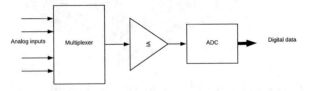

Fig. 5.5 Simple data acquisition system in IoT device

Signal conditioning circuits use op amp-based circuits like instrumentation amplifiers to pre-process the signals before they are fed to analog to digital converter (ADC) to generate digital data suitable for digital processing or direct feed to laptops or any other computing systems in IoT devices. The ADC is the last block between the analog and the digital signal path in such device architectures. In any sampled data system, such as a multiplexed data acquisition system, a sample-and-hold stage precedes the ADC to hold the input to ADC constant until the digitization is complete. This defines the acquisition time of ADC. Some ADCs have internal sample-and-hold circuits or use architectures that emulate the function of the sample-and-hold stage.

5.3 Electrical Isolation

Electrical isolation between sensor circuitry and the processing circuitry is very important in cases where:

- The voltage at the sensor circuit and the voltage at the processing circuit have a large difference.
- There are dangerously large-voltage high-tension wires or generators, large motors, turbines, welding machines, or machines using large current in the same building or on the same power network in IIoT scenario. Examples are laboratories, jet engine plants (power inverters), automotive plant, power plants, and test racks.
- There is a large fluctuation in the ground voltage.
- The power system is subject to electrical spikes or transients, which can cause fatal errors in systems in mission critical and safety critical applications like flight testing, Bio-medicals systems and in industry automation systems.

Electrical isolation translates or shifts measured signal property like frequency or timing to another circuitry without changing actual connection. Common examples are a transformer or an optocoupler.

5.3.1 Amplification

Sensor output signals are of low amplitude and hence are amplified using instrument amplifier. Instrument amplifiers are special amplifiers with the following characteristics:

- Very high open-loop gain
- Very low DC offset
- Very high common mode rejection ratio
- Low drift
- Low noise circuit

An example of commercially available instrument amplifier is INA333 and INA 326 from TI, an output rail-to-rail micro-power amplifier. Major design considerations of the signal conditioning amplifier are the sensor output voltage range and the input voltage range of ADC and, its reference voltage. The gain of the instrument amplifier is adjusted such that the total voltage swing at the input results in an output swing equal to the input range of ADC so that full range of digital representation is achieved. An instrument amplifier is an op amp-based circuit. The design guidelines for different instrument amplifier modules are different and have to be followed for proper functioning of the circuit.

5.3.2 Filtering

The input to ADC module in the data acquisition block of IoT is filtered for removing noise and optimized to convert it to digital signal for further processing. The successive approximation converter ADC compares the input voltage to the output of a DA converter for each digital number, thus sending the output digital code corresponding to the input signal. This conversion process hence requires stable input voltage for minimum time of conversion. The design of the *input filter*, a low-pass filter, at the input of ADC depends on (1) acquisition time of the ADC, (2) sampling ADC input capacitance, (3) time constant multiplier, and (4) full-scale input voltage. Acquisition time is the amount of time allowed to get the input voltage stored on the ADC input capacitor to the accuracy required by the ADC. The filter design is determining the resistance and capacitance (RC) component value to suit the ADC parameters. The RC component value of the filter is computed using experimental design guidelines. One guideline is that C value of the filter is selected to be 20–50 times larger than the input capacitance of the ADC. The resister value of the filter is determined by having RC values much less than 60% of the acquisition time of the ADC. The acquisition time is being multiplied by factor of 0.6 to give margin for op amp output load transient and small signal settling time.

That is, $0.6 \times$ acquisition time of ADC $\geq k \times R_{filter} \times C_{filter}$ where k is the time constant multiplier of the ADC.

Filter output can be connected to unity gain amplifier or inverting op amp circuits to get low CMRR and drive low input capacitance of ADC.

The input to an ADC may require filtering to avoid noise and spikes corrupting the input measured. A low-pass filter whose time constant is comparable to the acquisition time of the ADC is generally used. A normal rule of thumb is

$$0.6 \times \text{acquisition time of ADC} \geq k \times R_{filter} \times C_{filter}$$

where k is the time constant multiplier of the ADC.

Too large a time constant will result in the S/H not keeping step with the changing input.

5.3.2.1 Power Consumption of IoT

Power consumption of IoT device is an important performance parameter to work on in any applications and battery-operated ones in particular. Low power consumption is achieved by defining and implementing system modes of the device and turning them off when they are not active. RF functional blocks are generally the most power-consuming which have to be identified when non-functional and turned OFF. But it is essential to understand the latency of switching them ON when active again. This implementation uses identifying RF signal strength indicator (RSSI), and when it indicates no RF receive signal, the RF functional blocks are switched OFF by appropriate control circuitry. The controllable blocks can be enabled/disabled digitally through general-purpose input/output (GPIO) pins of the processor.

The instrument amplifier, isolation, and filter constitute the data acquisition block and when integrated with buffer amplifier and ADC constitute the signal conditioning circuit which generates the output in digital form for processing.

5.3.2.2 Storage in IoT Devices

As the sensor data from IoT is captured continuously, there is a huge requirement for storage within the device before transfer to cloud for big, structured data storage. The IoT storage device should be scalable, run and maintain firmware, provide low-latency access to data for processors, and be able to handle large enterprise-level workloads. More importantly, these storages should connect to different platforms as there can be many types of heterogeneous IoTs. This is achieved with super-fast non-volatile memory express (NVMe) storage protocol which is used in the super-fast solid-state drives (SSD) with PCI express (PCIe) interface. NVMe is a high-performance, NUMA (non-uniform memory access)-optimized, highly scalable storage protocol that connects the host to the memory subsystem. The protocol is relatively new, feature-rich, and designed from the ground up for non-volatile memory media (NAND and persistent memory) directly connected to CPU via PCIe interface. This is a new high-performance protocol standard storage which is very well suited for IoT device storage. Regardless of the size, NVMe directly communicates with the system processor as in Fig. 5.6 and works with all major operating systems.

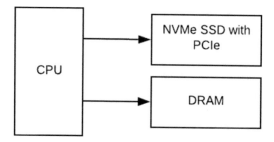

Fig. 5.6 NVMe memory over PCI express directly connected to CPU

Major features of NVMe protocol-based storages which suit IoT and IIoT are the following:

- It does not require custom device driver every time the software or OS changes.
- The memory organization can be planned to suit the power management strategy. Typically, memories consume more power when they are accessed for writing than reading. Certain memory lanes can thus be kept for writing so that the power optimization control can be added to them effectively implementing low power modes.
- NVMe protocol supports IO multipath, which is particularly useful for memory redundancy and load balancing purposes. If one path is not accessible or busy to CPU, data can be accessed via the other path.
- Effective partition of ready-to-read memory with initialization boot code can be done so that pre-OS firmware can be protected with secure boot applications.
- Asynchronous event capture is supported by the protocol like Self-Monitoring, Analysis, and Reporting Technology (SMART) status check, error reporting, firmware commits, and critical event management.
- Multi-stream writes are supported in NVMe protocol-based storage as shown in Fig. 5.7. Multiple independent IO transfers can be done simultaneously.
- To achieve high performance, certain part processing of the data is carried out in memory before they are written into the memory [2].

5.4   Hardware Accelerators

Hardware-software partitioning is the proven process in defining the architecture of any system design. This involves implementing time-critical functions of the system in hardware and flexible, slow process functions in software. *Hardware acceleration is a process of implementing part of the software-driven function or algorithm in hardware to take advantage of the speed, reduced latency, and other performance*

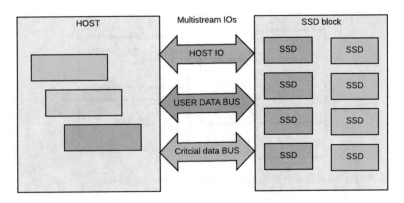

Fig. 5.7 Multiple streams of IO transaction in NVMe protocol-based SSD storage

advantages. This also improves overall efficiency of the function implemented. This is different from using a general-purpose processor for functional implementation in the sense that a part of the function is customized or feature/algorithm-specific function is implemented in hardware and software combination. For example, in a frame or packet processing function of the protocol processing, the frame/packet is required to be parsed into different fields, classified, type of packet is identified to send the, response packet in a communicating device. The packet processing function implementation can be done in many ways: i) completely hardware, ii) completely software, and iii) software with hardware acceleration. Part of function to identify different fields of the packet is implemented in hardware, and corresponding response action based on identified packet field is implemented in software. This combination of hardware-software circuit is called hardware accelerator. An example of a system architecture using hardware accelerator is shown in Fig. 5.8.

In the system architecture shown in Fig. 5.8, the hardware accelerator performs the packet processing function discussed in this section. The other functional blocks 1 and 2 and the processor subsystem perform functions as per the system partition defined in the architecture.

With the growing system complexity and increasing demand computational capability, it is more appropriate to use the hardware accelerators in the system design for increased performance and reliability. Examples of hardware acceleration include bit blit acceleration functionality in graphics processing units (GPUs), use of memristers in neural networks [2], and hardware acceleration for spam control servers [3].

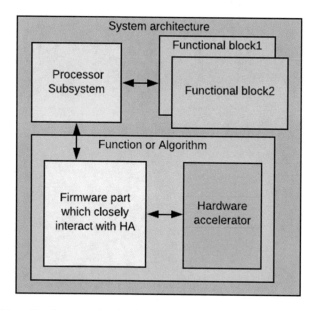

Fig. 5.8 Position of hardware acceleration in system architecture

In IoT or IIoT network, solutions are no different. The system latencies are the sum of device latency, gateway latencies, and cloud access latencies which are large. Applications like healthcare, autonomous cars, and industrial automation functions will not tolerate large latencies for critical functions. These functions are hence being shifted to edge devices closer to the IoT or IIoT device with high-performance system architectures for edge devices. As a thumb rule, the fixed computational resource of the system function or algorithm, which is reused many times in the functionality, is implemented in hardware accelerator, and the most flexible part of the function is implemented in software/firmware. For example, in a typical DSP filter design, the sum of product function is implemented in hardware accelerator, and the filter coefficient writing/reading is implemented in software.

Artificial intelligence and machine learning technology are integral part of IoT solutions. AI and ML require high-performance computing logic for training the machines with large data sets. This invariably uses hardware accelerators, which is called AI accelerator. In-memory computing is a new concept in hardware accelerator.

References

1. Data sheet of ADRF5046, Analog Devices
2. A Survey of ReRAM-based Architectures for Processing-in-memory and Neural Networks", S. Mittal, Machine Learning and Knowledge Extraction, 2018
3. High speed intrusion detection systems, packet classification, server load balancing, web-switching, bio-informatics, SANs, firewall load balancing and virus scanning" 17 July2014.

Chapter 6
Capturing Modules

6.1 Introduction

There are many applications which require real time or offline image or video processing functions. They need image and video capture modules. One can encounter many such applications in healthcare, agriculture, and forestry that demand live video monitoring. In healthcare, remote diagnostics, bedside care, and virtual clinics are the best examples. Another most interesting application which need video processing in health diagnostics is capsule endoscopy in which a capsule from Medtronic called "PillCam" is consumed which will take close to 50,000 pictures of the internal gut as it moves through it and transmit them wirelessly [1]. This is still in clinical trial stage, and once it succeeds, it can be a boon to cancer diagnostics and care. Precision and smart farming is another application which need video processing in agriculture. Some applications with image and video capturing and processing functions are shown in Fig. 6.1.

There are many applications for video surveillance systems in transport, automotive, and industrial domains. The IoT solutions are paving way in almost every field removing the redundant logistics of commuting using real time multi media processing.

6.1.1 Cameras

Cameras are the essential modules in applications involving image- and video-based IoT solutions. There are two types of image sensing technologies available: charge-coupled device (CCD) and CMOS. Both types of sensors convert light intensity of pixels forming an image into electrical signals. They provide better sensitivity and fidelity.

Fig. 6.1 (**a**) Smart farming (Courtesy: https://www.biz4intellia.com/blog/5-applications-of-iot-in-agriculture/). (**b**) Remote bedside healthcare (Courtesy: https://www.biz4intellia.com/blog/5-applications-of-iot-in-agriculture/)

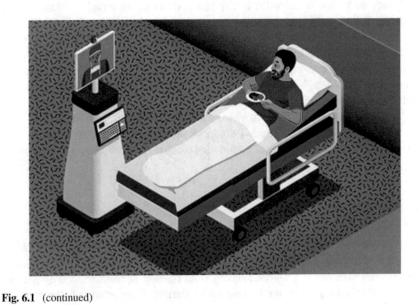

Fig. 6.1 (continued)

6.1.1.1 CCD Cameras

CCD cameras are a very old imaging technology; where some surface material exposure to light eject electrons, which are collected in a charge/potential bucket. CCD sensors collect charge in its pixels (basically, capacitive bins) and then physically shift the charge on the imager surface for sampling the output, which is an analog voltage where the charge is proportional to the light intensity. The charge is converted to voltage by a charge amplifier followed by an ADC for further processing. The whole image is exposed to the sensor simultaneously to all pixels for a fixed time. Figure 6.2 shows the block diagram of a CCD camera.

CCD chip is the array of photo-sensitive pixels which are organized as arrays with rows and columns, and is exposed to light. The term charge-coupled device actually refers to the method by which array of charges are moved around on the chip from the photocells to a shift register. Clock pulses create potential difference to move charges around on the chip, before conversion to voltage is done by a charge amplifier. The CCD sensor is an analog device, but the output is converted to a digital signal by means of an analog-to-digital converter (ADC) in digital cameras. The resultant stream of pixel data is stored in a large video buffer memory. The microcontroller in device will process the buffer content and transfer it to the destination on the Internet. Core of the CCD camera is the pixel array in the CCD chip. CCD image sensors are most used where high dynamic range and sensitivity are needed as in digital astrophotography and visual machine inspection.

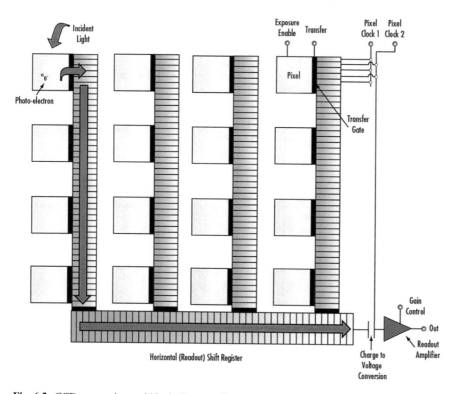

Fig. 6.2 CCD camera internal block diagram. (Image courtesy: Edmund Optics)

The CCD sensor is limited by the charge transfer speed but is highly sensitive and has a very good pixel-to-pixel consistency. Some of the main properties of CCD camera modules are the following:

1. module is analog device and signal processing is done outside the sensor chip
2. slow in speed as data is transferred pixel by pixel
3. High dynamic range and consistency
4. has Low noise and does not show moving shutter effect, a phenomenon which causes image to unnaturally wobble or shake
5. Consumes higher power and requires typically 7–10 V supply

6.1.1.2 CMOS Cameras

CMOS image sensors convert charge to voltage at the photocell or pixel itself, and the signal is multiplexed by row and column to multiple on-chip digital-to-analog converters (DACs). Inherent to its design, CMOS sensor is a digital device. Each photo site is essentially a photodiode and three transistors, setting and resetting or the pixel, amplifying, converting charge to voltage and multiplexing (Fig. 6.3). CMOS sensors are characterized by high speed, low sensitivity and rolling shutter noise in images due to spatial variation in fabrication process in pixel rows. This is

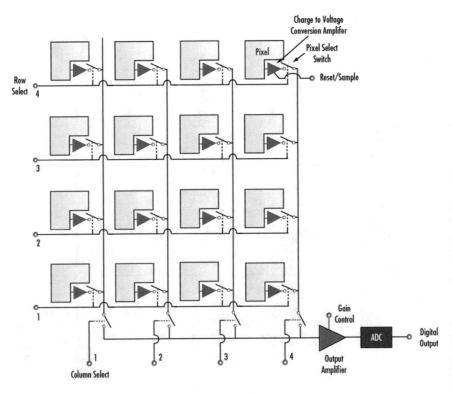

Fig. 6.3 CMOS Image sensor. (Image courtesy: Edmund Optics)

Table 6.1 Comparison of CCD image sensor and CMOS image sensor [2]

Sensor	CCD	CMOS
Pixel signal	Electron packet	Voltage
Chip signal	Analog	Digital
Fill factor	High	Moderate
Responsivity	Moderate	Moderate–high
Noise level	Low	Moderate–high
Dynamic range	High	Moderate
Uniformity	High	Low
Resolution	Low–high	Low–high
Speed	Moderate–high	High
Power consumption	Moderate–high	Low
Complexity	Low	Moderate
Cost	Moderate	Moderate

overcome by parallel processing of the pixels at the cost of silicon space. This has the advantage of very low power since it is VLSI fabrication. These are smaller than corresponding CCD camera counterpart for the same pixel array size.

A table of comparison of CCD and CMOS sensors is given in Table 6.1.

Major criteria to assess the performance of the camera modules for an application are the following:

- Pixel size: pixel array size…(nXm) of the image. Larger the pixel better the image.
- Quantum efficiency: Percentage of photons converted to electrons at a particular wavelength. Higher value is better.
- Frame rate: The frame rate refers to the number of full frames composed in a second. For example, an analog camera with a frame rate of 30 frames/s contains two 1/60 s fields. In high-speed applications, it is beneficial to choose a faster frame rate to acquire more images of the object as it moves through the field of view (FOV).
- Camera size: The size of a active area of camera sensor is important in determining the system's field of view (FOV). Given a fixed primary magnification (determined by the imaging lens), larger sensors yield greater FOVs. There are several standard area-scan sensor sizes, 1/4", 1/3", 1/2", 1/1.8", 2/3", 1", and 1.2", with larger available (Fig. 6.4).
- Temporal dark noise: Noise in the sensor when there is no signal.
- Saturation capacity: Amount of charge that a pixel can hold.
- Maximum signal-to-noise ratio: The highest possible signal-to-noise ratio including shot noise and temporal dark noise is good for sensor.
- Absolute sensitivity threshold: Number of photons needed to have signal equal to noise.
- Gain: Parameter indicating change in a number of electrons needed to observe a change of 1 bit in 16-bit ADUs (better known as grayscale).

Image sensors are available by different vendors as ready-to-integrate modules for IoT applications. The module contains the image sensor, ADC, and image buffer memory which can be accessed serially or in paralle interfaces

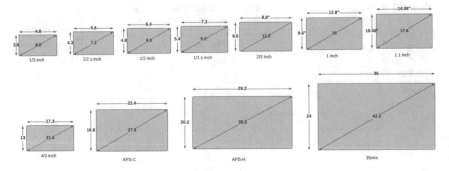

Fig. 6.4 Camera size

6.2   Smartphone Camera Modules

Smart phones are equipped with cameras which produce very-high-quality images at a much faster speed. This is achieved by advanced image video processing techniques adopted in addition to good-quality cameras. This module resides in the smartphone chipset/CPU on which most of the advanced algorithms run providing additional enhancements and special effects both while capturing images and later. Some of the algorithms and functions available are face recognition, filters, panoramic scene capturing, and object identification.

Images are also geo-tagged with the GPS coordinates if the phone has an internal GPS chipset.

6.2.1   Specifications of Camera Modules

Important functional specifications of camera modules are:

Pixel Size: The resolution of the image taken by a smartphone is measured in megapixels. Higher number of pixels may not always result in sharper images as the quality of the image also depends on the resolution of the CMOS image sensor used. A large number of pixels, however, help crop the captured image by retaining most of the image features. In most cases, still images are stored as either jpeg (Joint Photographic Experts Group) or HEVC (High Efficiency Video Coding) for videos, which compress the image file size without loss of detail (loss-less compression). The most common format for recording video is H.264/H.265. Some high-end phones capture in RAW leading to much larger file sizes.

Aperture: The aperture of a lens indicates how much light the lens lets in. The larger the aperture, the more light is let in; conversely, a smaller aperture lets in less light. Aperture is measured in f-stops. The larger the aperture, the lower the f-stop number. The quality of the optical system is determined by f-stop number.

Electronic Image Stabilization (EIS) and Optical Image Stabilization (OIS): Helps eliminate camera shake and produce a better-quality image. Some phones successfully use digital OIS, but the good phones use a mechanical optical image stabilization system.

Autofocus: Modern smartphones have autofocused system fitted near the camera lens system. This helps automatically adjust the focus looking at the distance of the subjects.

Phase-Detection Autofocus (PDAF): PDAF technology uses paired masked pixels on the image sensor, designed to mimic your eyes. The image signal processor (ISP) adjusts the camera lenses until both images are synchronized with each other indicating that the subject is in focus.

High dynamic range (HDR) attempts to add more "dynamic range" to the images. As opposed to taking one photo, HDR mode takes three photos at different exposures. The CPU/ISP combines these three images and highlights the best parts of each photo.

6.2.1.1   Videos

The majority of smartphones are capable of capturing minimum of 720p HD video 30 frames per second (FPS). The higher the FPS, the smoother the video or, in the case of slow-motion shooting, the slower you can make a video without losing quality. The higher the resolution and the higher the number of frames, the more storage space is required. **Standard** and **slow-motion** captures are two video formats which are common. Standard format has a resolution of (1) 1080p HD with 30 and 60 FPS and (2) 4 K at 24, 30, and 60 FPS. Slow-motion captures have the resolution of (1) 720p HD with 240 FPS and 960 FPS and (2) 1080p HD with 120 FPS, 240 FPS, and 960 FPS.

6.2.2   Commercial Image Processing Modules

System on chips (SOCs) with 1.3M pixel image processing modules with sensor, internal memory and ISP, which can run advanced DSP algorithms for edge detection, optical flow analytics and easy interface to digital processors make high quality advanced image processing systems. Following are few of the well known high performance sensor modules:

6.2.2.1   STMicroelectronics VG*640: 1.3 Megapixel High-Dynamic-Range Image Sensor

VG *640 is a high-performance, high-dynamic-range 1.3-megapixel image sensor. Designed for automotive, security, and other demanding outdoor applications, the device offers good low light performance and many safety integrity features.

Fig. 6.5 Raspberry Pi camera modules. (Source: www.raspberrypi.org)

6.2.2.2   Raspberry Pi Camera Module V2(SKU)

Raspberry Pi camera [3] shown in Fig. 6.5 developed by the Raspberry Pi Foundation
comes in a variety of pixel densities. There are two types of camera modules: standard camera module which can take images in white light and NoIR camera modules which will not have IR filter and can take photos in the dark. The cameras come
with different Sony IMX image sensors with different resolutions and performances
which define the image quality. The modules are compatible with most of single-
board computers (SBC) or development boards which have a standard CSI (Camera
Serial Interface) or parallel interface. The modules are supported by the APIs in the
Raspberry development IDEs for application development.

6.2.3   Camera Modules for Arduino

Camera modules like ESP32-CAM can be interfaced with Arduino boards through
Serial Camera Control Bus (SCCB) interface which is compatible to I2C interface.

References

1. www.medtronic.com
2. Edmund Optics: Imaging electronics: Understanding camera sensors for machine vision
   applications
3. www.respberrypi.org

Chapter 7
IoT Software Design Methodologies

7.1 Introduction to IoT Software

IoT and IIoT solutions need different software at different component levels of the system framework. The IoT devices are embedded systems with sensors where the software will be embedded in the system memory. This software is called *embedded system software or firmware.* The remote system gateway to which the IoT devices are connected will have *data computation software.* The processed data from sensor device is further communicated to the cloud on the Internet or through *gateways* for analytics, visualization, and AI-based applications using *cloud software.* The security aspects have to be plugged into both device software and application data while in storage and during communication in the entire framework. The IoT device architecture with different types of software running in them is shown in Fig. 7.1.

In Fig. 7.1, the boxes with dotted outline indicate options available for supporting the functions. Boxes shown in orange indicate different intelligent algorithms whether it is device intelligence, gateway intelligence, or edge intelligence supported depending on the IoT application. Boxes in pink indicate the level of security implemented. The security levels can be of many types like: it is flash encryption (for securing firmware) or securing boot loader or Internet security. The boxes in blue indicate different protocols supported to have web connectivity to access cloud server. The boxes in red indicate the connectivity options supported for communication, whether it is Bluetooth [1], Wi-Fi, LoRa [2], or cellular. Firmware upgrade is the feature very much needed for IoT solution for regular upgrading by feature addition or bug fixes. IoT device can implement mix of these combinations depending on the application. Each of these functional boxes needs software developments using different development platforms corresponding to signal processing, data computation electronics, communication electronics, protocol implementation, security, adding intelligence, and server systems. When there are so many options

© The Author(s), under exclusive license to Springer Nature Switzerland AG 2021
V. S. Chakravarthi, *Internet of Things and M2M Communication Technologies,*
https://doi.org/10.1007/978-3-030-79272-5_7

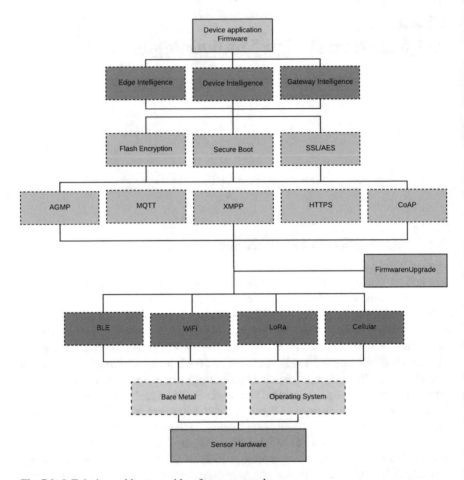

Fig. 7.1 IoT device architecture with software protocols

for the device architecture there are many considerations to chose the software development options for IoT designs. They are discussed in following sections.

7.2   Selection of Hardware

Embedded firmware development required primarily depend on the choice of hardware. Selection of hardware is predominantly the choice of microcontroller. Choice depends on the processing power needed for the application and the interfaces it supports for sensors and actuators. Since the microcontroller subsystem includes on-chip memories, choice depends on whether the application demands operating system-based architecture or bare metal architecture (application software without the operating system).

7.3 Operating System or Bare Metal Architecture

This is one of the important design considerations in IoT software development. Choosing operating system is to ease the firmware development and to reduce the development time. Operating system (OS) takes care of scheduling, multitasking, and multithreading various functions in the software but increases the software size and hence the memory requirement in the device. Nevertheless, for simple device functionality, it is good to avoid an operating system reducing the software size (footprint) and hence the unnecessary design overhead. However, development time increases, but of late, the chip vendors are providing ready-to-use library module to address this as a part of the integrated development environment (IDE).

7.3.1 IoT Software Footprint

It is always a good practice to minimize the software footprint as the device software resides in on-chip flash or on-chip RAM which directly impact device performance like power, speed, and size. Commonly used low-cost microcontrollers typically have 2 MB on-chip flash, which needs to be considered as it has to store the code as well as data captured by the sensor before being transmitted to other system components in IoT framework.

7.3.2 Communication Device

Implementing communication functionality is another important design decision in IoT device as it connects to the Internet. As shown in Fig. 7.1, there are many ways to access the either through Internet gateway or M2M devices. Latency is the key factor in deciding the gateway requirement for communication and the level of intelligence to be adopted at the device, the gateway, and the cloud in the framework. This is a major design decision in the IoT system architecture and accordingly firmware architecture. Also, the decision on where the communication software should be implemented whether in the IoT microcontroller or a separate dedicated microcontroller has to be made depending on processor resources like MIPS and on-chip memory.

7.3.3 Power Management

Most of the IoT devices operate on battery and hence power management is very important even in software architecture. One may not get total power optimization if the software does not effectively use power management modes supported in most of today's microcontrollers. Software needs to understand the system requirement and low power modes (sleep modes, hibernate) supported in the hardware to get advantage of power optimization of the complete system.

7.3.4 Firmware Upgrade

Firmware upgrade ensures that the firmware may be changed either for added feature or for bug fixes as workaround a bug to device when it is in field use. If the IoT devices have wireless communication capability, the firmware upgrade is done over the air (OTA) remotely through wireless technologies like Bluetooth, Wi-Fi, Zigbee, LoRa, etc.

7.3.5 IoT Security

Security in IoT is an important feature as subsystems are distributed in multiple platforms, devices, gateways, and cloud servers. As the number of connected devices increases, security threat also increases. This calls for implementing security in firmware designs, storage and communication. Security is absolute necessary even in low-cost devices.

The development platforms for the software design range from embedded development platforms to cloud server-based software development platforms. The programming languages typically used are, C, embedded C, C++, JavaScripts, Pythons, and HTML and each require different kinds of skill set to design and develop IoT system solutions.

7.4 IoT Embedded System Software

IoT device electronics, which capture and collect data from sensors and control external devices through actuator circuits, form embedded systems. The embedded system software used in IoT devices are developed for a specific hardware. It is also called firmware as such software is fixed or firm and need procedure to upload newer versions of the firmware. The firmware for the embedded system turns ON the system as. soon as it is powered using **boot loader, collects data** from sensors, provides **external control** through the actuator, **stores** the collected device data, and **enables the communication interface** for data transfer.

The firmware also enables/disables different set of configurations of the microcontroller unit (MCU) and other circuitry depending on the functional modes defined in the architecture. The boot loader and firmware can be downloaded onto the embedded memory like flash memory or EPROMs for the device to work independently.

Firmware is developed in special environment called integrated development environment (IDE) designed to support development of functional modules using microcontrollers, IoT sensors, actuators, storage, and computation to an extent. IDE provides access to device-specific web and software development kit (SDK), tools software and hardware configuration tools. Integrated development environment (IDE) featuring industry-standard code editors, compilers, and debuggers; and advanced value-added tools for network analysis, deriving code-correlated matrices, and energy profiling is used for development of firmware. There are many IDEs which run on laptops, computers, workstations, or cloud server. For example, the *Simplicity Studio* is a development platform for Silicon Lab's MCU-based IoT hardware kit. There are online *embedded development platform*, which run on is a cloud-like *Mbed OS* for ARM Cortex-based microcontroller. *Open-source Arduino IDE* is another IoT software development environment. All these platforms are free for limited use available with user registrations and support standard interface devices with their driver module in their library. These driver modules are software, which operate on the hardware when interfaced with the chosen MCU-based boards. Development platforms Mbed OS and Simplicity IDE also support Arduino interface as there are many devices supported in Arduino environment and its popularity.

7.4.1 Embedded Software Development Platforms

7.4.1.1 Arduino IDE

Arduino IDE is a very popular integrated development environment that supports embedded software development for a range of low-cost Arduino boards based on AVR microcontrollers. Popular Arduino boards are Arduino Uno, Arduino Nano, Arduino Micro, Arduino Leonardo, and Arduino Nano entry. More on this can be referred in website https://www.arduino.com. Typical IDE is shown in Fig. 7.2a and 7.2b.

Arduino IDE helps develop application-specific software on the family of Arduino boards for the target application. The IDE supports **text editors** for writing programs, a message window where problems in the code are displayed for programmer interaction, and *tool bar* with menus and buttons for commonly used quick functions, like copy, paste, search, and replace, indentations, formatting the code, etc. It also supports *program explorer* displaying the directory structure and program files in the directories and *object code explorer* to traverse to look for code objects and *simulator* for functional verification of the program. The IDE can connect to the hardware boards to upload the developed program by selecting target hardware board in "hw board selector" and connecting USB port of the computer where IDE is executed to the USB port on the embedded kit. IDE board USB port. The text editor in IDE has program friendly features like color coding, syntax highlights, function exploring, and hardware viewing. There are a family of Arduino hardware boards available for IoT devices, a few of

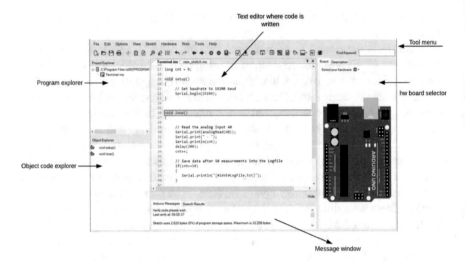

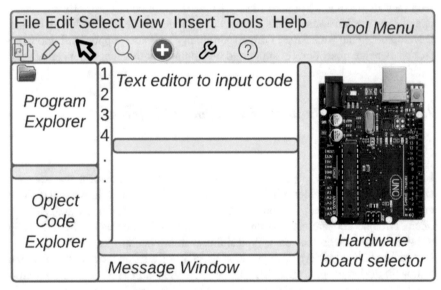

Fig. 7.2 Arduino IDE (Courtesy: www.arduino.com)

which are shown in Fig. 7.2. There are many small daughter boards with specific functions developed by many vendors which can be interfaced with any of the Arduino boards shown in Fig. 7.3 to enhance the functionality like extra memory mkr_mem_shield, sensor functions, Wi-Fi connectivity Uno_wifi_connectivity, Bluetooth connectivity Arduino Nano Ble, Mkr_can_shield, etc.

The program in the Arduino IDE is called *sketch*. These sketches are written in programming languages supported by IDE and saved as files with specific file extensions. Arduino IDE supports C, C++, and some advanced languages like HTML, HTML 5, JavaScript, etc. There are other alternatives for Arduino IDE with additional

Fig. 7.3 Arduino IoT boards. (Courtesy: www.arduino.com)

programmer-friendly features. Most open-source device drivers for the add-on device modules are called *shields*. These shields are available as library modules in the Arduino IDE. One or more shields can be imported or included in the user program depending on required functions for target application. For example, if the IoT device requires M2M communication using Bluetooth, the Bluetooth shield corresponding to BLE daughter board "Nano Ble" can be included along with the basic computing embedded software, and device embedded program can be developed.

7.4.1.2   Simulator

Function Simulator is a software tool in the integrated development environment which enables the programmer to verify the functionality of the program without the actual hardware. The simulator permits the programmer to develop the test environment where required stimulus for the function can be fed into the program from the test bench and the response can be observed through waveforms. Once the functionality of the program is verified on the simulator, the software can be loaded onto the hardware.

7.4.2   Embedded Program for an Application

The embedded system program for an IoT device consists of the following.

7.4.2.1   Device Boot Loader

The device boot loader is that part of the program which brings the hardware to active state once power is turned ON. This is the first program to execute for any microcontroller. It initializes the program counter to the address of the next instruction to be executed by MCU. The program counter loads the operating system from external sources and initializes control to the user program on the module or externally when the board is turned ON. Device booting is deemed complete when the user interface is enabled or normal functional screen is displayed for the IoT.

7.4.2.2   Operating Systems

The real-time operating system (OS) enables the use of on-board features by properly scheduling different functional tasks and instructions. The OS residing in external memory or flash, gets loaded to internal RAM for easy execution and activates the user screen for user interaction or normal functional mode of the device. Operating system (OS) controls workflow of different tasks of the program by properly managing the sequence of instructions, tasks, and processes. It also allocates memory to these tasks, as appropriate, for proper execution by the microcontroller. When the OS is capable of processing events in real time, it is called real-time operating system (RTOS). Events can be from the processor, IO devices, the user, or any functions or processes.

7.4.2.3   Hardware Configurations and Initializations Setup

Hardware configurations and initializations are important to make the device start functioning from known functional states. Typical initialization includes configuring data transmission rates, default filter co-efficients, default input-output modes for general-purpose IOs, default switch position controls, transmission power configurations, gain factors, etc. These are done by writing to the configuration registers in hardware. This will ensure the hardware comes up in a known default state when powered up or reset.

Signal Processing Algorithms

These are application-specific functions needed for processing or conditioning the signals captured by sensors for improving the SNR and reliability before further processing by the digital microcontroller. These are typically digital noise filters, frequency converters, and other DSP or RISC algorithms.

Computation Functions

These are programs for computation functions targeted to microcontroller units in the hardware. The microcontrollers have computational blocks, enhanced input-outputs, PWM functions, memory functions with internal RAM, flash, GPIOs, serial interface functionalities, port functions, and timers. Advanced microcontrollers even have functions like signal converters like ADC/DACs, comparators, advanced arithmetic functions like complex number processing, etc. Figure 7.4 shows the functional blocks of ATmega32 AVR microcontroller which is used in Arduino boards.

Interface Module Configurations

The functional modules interfaced to the MCU hardware board are sensors, actuators, and other blocks like ADC and DAC modules. These blocks are interfaced to MCU either through GPIOs or by standard interfaces like I2C or SPI. These

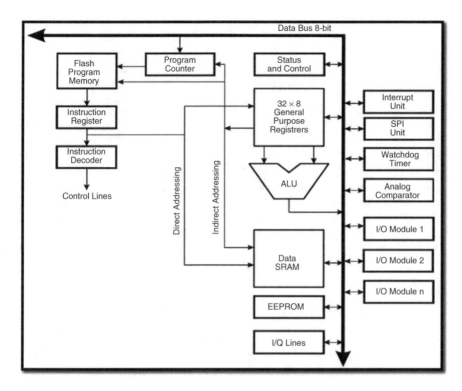

Fig. 7.4 Functional block diagram of ATmega32 AVR microcontroller from data sheet (Courtesy: Atmel)

interfaces generally need programs for driving interface signals complying with the relevant protocol to write to configuration registers or reading the status registers inside the modules. This function is part of the embedded programs.

Communication Functions

The communication shields need codes to configure interface communication modules connected to the MCU for transmission and reception of the processed data to/from other machines.

Communication Functions

The communication shields need software to configure interfaced communication modules connected to the MCU hardware for transmission and reception of processed data to/from other machines.

Figure 7.5 embedded firmware code snippet of program in a design case which captures values from DHT11 sensor.

```
#include <dht.h> // Including DHT shield corresponding to DH11 temperature-humidity sensor

#define dht_apin A0 // Analog Pin sensor is connected to

dht DHT;

// System initialisation and configuration

void setup(){

  Serial.begin(9600); // configuring serial interface to baud rate of 9600 bit per second
  delay(500);//Initialization of delay for system boot up
  Serial.println("DHT11 Humidity & temperature Sensor\n\n");// Printing the text for user
  delay(1000);//Dealy for sensor to capture temperature and humidity before it is read by
Controller.

}//end "setup()"

//Interface functions to read sensed data from sensor through SPI serial interface

void loop(){
  //Start of Program

  DHT.read11(dht_apin);

  Serial.print("Current humidity = ");
  Serial.print(DHT.humidity);
  Serial.print("% ");
  Serial.print("temperature = ");
  Serial.print(DHT.temperature);
  Serial.println("C ");

  delay (5000); //Wait 5 seconds before accessing sensor again as sensor can read data once in
2 Seconds,
                    // extra safe delay is added before fetching subsequent sensor data
                    //Fastest should be once every two seconds as per the specification of Sensor
}// end loop
```

Fig. 7.5 Arduino program to interface DH11 sensor

As one can see, the entire embedded program can be classified into two parts: **i)** **setup** and **ii) loop** functions. The **setup** part of the sketch in Arduino deals with providing the one-time pin connectivity information, making the GP input or output. The **loop** part of the program consists of code which gets executed continuously. The loop part of the program contains codes for all the required application-specific functionalities which are executed continuously like reading of the sensor data and waiting for set delays as shown in an example program shown in Fig. 7.5.

7.5 Embedded Software Architecture Guidelines

It is necessary to develop an IoT embedded software with good practices. The life cycle of the device development depends on the architecture of the software developed. Therefore, it is necessary to understand the guidelines for good architecture of firmware with a few important parameters:

1. *Modularity:* Modularity is important in firmware architectures. Instead of a firmware that has several thousand lines of code with no functional boundaries, which is very difficult to maintain, one should have a clear functional description and interface, with clarity of inputs and outputs, with functional modules in the firmware application. This makes the modules reusable across generations of similar products minimizing the development time. In most cases, it is possible to reuse the modules if they are modular to develop a new product. The functions should have clear functional description and interfaces in a module which can be copied and connected to other compatible dependent blocks in the firmware design as shown in Fig. 7.6 with arrows showing clear dependency with respect to each other. Module 2 and module 3 depend on the output of module 1, and module 4 depends on the outputs from module 2 and module 3.

   Dependencies could be either compilation or execution dependency. The compilation dependency occurs if the variable used in module 1 is declared in module 2 or module 3 as shown in Fig. 7.7 in which case module 1 will not be compiled unless modules 2 and 3 are compiled.

   Execution dependency is straightforward when the function in module 2 depends on the output of module 1. Imagine the firmware application involving thousands of functions dependent on each other; a dependency diagram will be like a spaghetti diagram. Fixing issues in such software will be a herculean task leading to a large development time. Another problem in such software is loop dependency, which will lead to large code memory occupancy in hardware. As a guideline, it is essential to keep dependencies between modules to a bare minimum and totally avoid loop dependencies for efficient debug, maintainability, and reuse.

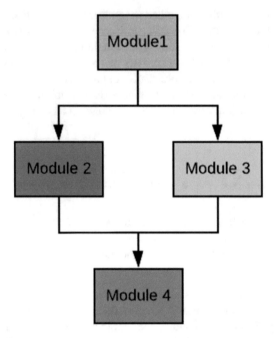

Fig. 7.6 Modular firmware architecture

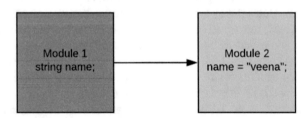

Fig. 7.7 Module dependency for compilation

2. *Encapsulation:* Like in object-oriented programming (OOP) , internal functionality of the module has to be made private to the module. The internal operations of the module are not exposed or accessible to other modules. This ensures that any change inside the module is limited to only that module. This is achieved by defining classes in the functions with most logical set and get interfaces. This methodology is called application programming interface (API). This is demonstrated in Fig. 7.8. In the example shown, the module 2 need not know the variable of *my_name* which is internal to module 1; instead, it will get *ASHA* as a return API call from *Get_name.*
3. *Upgradability:* Good architecture should support addition of new features as add-on functions. This is essential as new IoT devices are not designed from

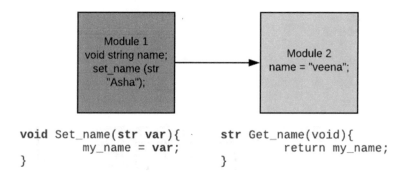

```
void Set_name(str var){           str Get_name(void){
          my_name = var;                      return my_name;
}                                 }
```

Fig. 7.8 API-based module functional calls ensuring encapsulation

scratch but built on existing functional blocks by upgrading with additional features. For example, the Apple watch initially did not support call functions,
which later versions supported by firmware upgradation.

4. *Scalability and Portability:* These are the two major requirements when it comes
   to IoT solutions. Scalability ensures newer devices for replacement or upgradation as a regular feature. This should be kept in mind when firmware architecture
   is being defined. Portability to newer powerful hardware is also a regular requirement in IoT solution. This is achieved by defining a hardware abstraction layer
   (HAL), which is hardware dependent, but all other functional layers are independent to HAL. When the hardware changes, only HAL is updated keeping all
   other modules the same with minimal changeover time as shown in Fig. 7.9.
5. *Testability and Verifiability:* Architecture should be test and verification friendly.
   This means each module, which forms a feature for the device requirements,
   should be verifiable independently and at integrated level. The architecture
   should also be suitable for regression testing using simulators and emulators and
   with real hardware.
6. *Documentation:* Architecture should be well documented with message charts,
   event or timing diagrams, flowcharts, state machines, class diagrams, etc. with
   the right level of expectations with respect to coverage matrices.

A good IoT firmware architecture makes the system reliable, robust, low-power,
secure device operation.

7.5.1   Embedding the Software onto Hardware

Embedding the software onto hardware is uploading the software developed in
embedded IDE onto hardware. The IDE can access the information from hardware
board through physical connection called programming cable connected between
the computer where IDE is installed and the hardware board: USB cable in the case
of Arduino IDE and Arduino boards. It is also the means by which the software is

Fig. 7.9 Hardware
minimally dependent FW
architecture

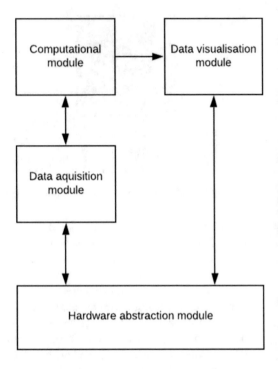

uploaded onto hardware. It enables reading of messages from the embedded software located in flash memory in hardware board into the computer screen where IDE is set up. These are messages during testing and debugging the downloaded software. In Arduino, it is accessed through serial **COM** port on computer. The Arduino board is listed as one of the ports on the IDE and software developed (is compiled using **Build** command in the IDE), and the built sketch is uploaded onto the hardware using Build button on the IDE tools.

7.5.2   RTOS and Mbed OS

The embedded software for IoT may need real-time operating systems when a number of functions in the applications are complex and are in large number. The operating system takes care of prioritizing and scheduling multiple functions called multiple threads for execution by the microcontroller efficiently and without overlap or overheads. There are many operating systems which can be used for IoT device and communication software. Linux OS, ThreadX, VxWorks, FreeRTOS, DuinOS, and Arduino versions real-time device operating systems are available for Arduino-based boards, which support multitasking of programs. **Mbed OS** is a free open-source operating system which helps development of IoT device software targeted to the Mbed-enabled hardware platforms or boards. These are ARM Cortex M4 microcontroller-based boards. IoT embedded software for ARM-based micro-controllers can be developed in this environment. Mbed OS also supports Arduino interface for interoperability and expansion of device features for Arduino boards when they are used with ARM Cortex M4-based IoT boards. It has all the modules of any other IDE like code editor, functional libraries, networking libraries, RTOS-based application development, build tools, test and debug scripts, simulator, debug-ger, compiler, message viewer, and device support feature for uploading onto the hardware. It is an online platform for the development of embedded software for IoTs.

7.6 Most Used IDEs

Every microcontroller manufacturer like Intel and AMD, offers their own develop-ment platforms for the embedded software development. Intel Galileo is IDE for Intel microcontroller-based Galileo boards. Embedded IoT application is developed for Intel Atom® processor and Intel Quark™ SoC-based hardware using *Intel systems studio* with *Useful Packages and Modules* (UPM) that support sensor/actuator drivers using MRAA input/output library interface. Libmraa is a C/C++ library with bindings to Java, Python, and JavaScript to interface with the I/O on Intel's Galileo, Edison, and other platforms. The APIs are available in both C++ and JAVA.

Integrated development environment (IDE) software like MPLAB® X IDE and Atmel Studio are used to develop firmware for PIC and AVR chipset-based hard-ware IoTs. It also supports SDKs for many sensors, Internet, and other M2M com-munication modules.

Simplicity Studio is a development platform for Silicon Labs portfolio of boards EFM8, EFM32, EFR32, EM35x, and EFP (www.silab.com).

All these IDEs support Arduino shields for interoperability with Arduino boards. They also support operating system-based embedded software development. And these platforms also host a variety of device drivers in the library which can be just included in the device program and used. Embedded development platform is a combination of different tools like professional code editors, functional simulators, debuggers, and flash programmers and includes many library modules.

Integrated development environment (IDE) Atollic TrueSTUDIO®, Keil® µVision, NXP CodeWarrior™, SEGGER Embedded Studio®, Atmel Studio, TI Code Composer Studio™, Wind River Tornado®, Wind River® Workbench, etc., the development frameworks like Microchip Advanced Software Framework (ASF) and Atmel START and STM32CubeMX are used for firmware development for supported hardware. IDE support firmware development for Internet connectivity functions of IoT chipsets.

7.7   Software Development for Communication

IoT and IIoT devices should be able to access the Internet. Devices can access the Internet directly through Internet connectivity or through edge devices and gateways to cloud connectivity. Right choice of communication strategy is very important for getting best performance of the IoT systems. The more proprietary the communication protocols, the more difficult it is to connect the devices to the Internet for interoperability. The more standard the communication protocols used by devices, the better their extensibility and flexibile to connect to other devices. Hence, firmware designers need to keep this in mind during development of the connected device. Suitability of connectivity solutions needs to be considered during the IoT system design phase itself. There are several low-power protocols being defined specifically for device connectivity. There can be many IoT and IIoT architectural solutions depending on the applications with different combinations of connectivity out of them; three architectural solutions are shown in Fig. 7.10. Figure 7.10a shows IoT healthcare system solution where all the health monitoring and diagnostic medtech IoT devices are connected to the virtual private cloud server in the hospital. This ensures data privacy and security to hospital-subscribed users. Figure 7.10b shows a typical IoT solution for automotive applications where the response time is critical and hence uses edge computing configuration. Figure 7.10c shows an industry automation solution with IIoT with edge computing where response time to control the machine parameter is critical and is not required to bear the overhead of communicating to cloud for actuator controls. The analytics in all three cases can still be run by the cloud server with Internet connectivity.

Short-distance or long-distance connectivity within each of the solutions is to be rightly chosen depending on the range at which the IoT devices are deployed and where the data storage, analytics, and control are required. For short-range communication, BLE-, Zigbee-, and Wi-Fi-based connectivity are chosen, while for long-range communication, cellular (LTE)/LoRaWAN [2] technologies are considered. Many vendors offer cloud computing resources for processing captured and gathered IoT data in platform as a service (PAAS) model of engagement [3]. Most used ones are AWS® IoT, Azure® IoT Hub, Google Cloud™ IoT, IBM Watson™ IoT, etc. It is easier to connect devices to these platforms considering the kind of support guaranteed from these cloud service providers.

There are IoT applications like healthcare and automotive where the latencies of data transfer to cloud and response times cannot be tolerated. These applications demand faster response times for the actions to be taken based on the data collected and analyzed. For example, for crash management of vehicles in autonomous cars, the action has to be immediate. For such applications, where the response latency is critical, the intermediatory edge functionality is created where the all the cloud resources are available at the edge device. In these systems, the sensor data is communicated, stored, and analyzed in edge device, and immediate response is generated in real time. Even for industrial automation system shown in Fig. 5.10c, data from IIoTs need not be transferred to cloud servers for control action but can be

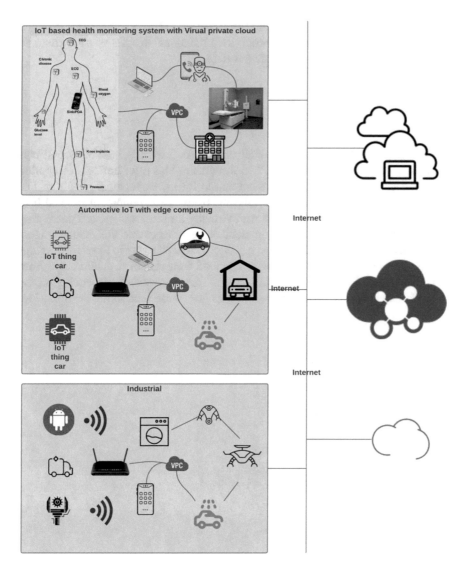

Fig. 7.10 IoT and IIoT system solution configurations

processed locally in real time without suffering the latencies of the Internet and cloud processing. Faster response for controlling the IoT and IIoT device is received from edge devices.

The embedded device IDEs support IoT device software development of programs for both short-distance edge and local VPN and Internet connectivity to public cloud. Depending on what connectivity technology is chosen, respective SDKs are to be downloaded and included in the device software. For example, Arduino IDE supports Ethernet library and header files for Ethernet shield, Ethernet client,

Ethernet server, domain name system (DNS), Dynamic Host Configuration Protocol (DHCP), and User Datagram Protocol (UDP) functions. Respective APIs are to be used to establish connection and data transfers.

7.8   Cloud Computing

Cloud computing technology is the main enabler for flexibility, scalability, and anytime access of the IoT devices and data from them. This offer limitless computing and storage resources. This is the main reason for the exponential growth of IoT devices in many domains. The data collected from a large number of IoT and IIoT devices can be stored and analyzed with virtually limitless storage and computing resources using cloud computing on cloud servers. To enable this, many vendors offer cloud resources as "platform as a service (PAAS)" to the IoT system developers. A few well-known vendors are AWS cloud from Amazon, Azure cloud from Microsoft, and Google Cloud from Google. They are scalable depending on the solution requirements. Main service offerings from these vendors are the following:

- Expandable cloud storage
- Device connectivity to cloud service
- Docker service
- Data security
- Database management system
- Big data analytics
- Machine learning tools
- Analytical development tools
- Solution developer tools and scripts
- Management tools
- Access through web and mobile applications

7.8.1   Expandable Storage

Expandable high-performance storage is an object storage service that is scalable and secure. Any IoT or IIoT solution targeted to any application in any domain can use this service to securely store the data. The data can be transferred to this storage from mobile applications, websites with backups, and archive services for big data analytics services. Data can be stored to this memory service from all the subscribers all around the world.

7.8.2 Device Connectivity to Cloud Service

IoT devices can easily be connected to cloud without the need to learn about provisioning or managing the data servers. Cloud can support billions of IoT or IIoT devices and trillions of messages and can process and route those messages to other devices reliably and securely. With this service, IoT applications can keep track of and communicate with all the connected devices, anytime, anywhere. The cloud-connected device can also access other services offered by the cloud server provider like analytics, storage, machine learning algorithms, etc.

7.8.3 User Access Management

The IoT systems enable many business models, and one of them is services around IoT system. A simple example is monitoring and maintaining the clean rooms in IC fabrication houses. This requires giving access rights to different users for the selected services. The user access right management service on cloud enables registering the user and managing the access rights to various other cloud services for them.

7.8.4 Data Analytics

IoTs generate a large set data, which are automatically stored in cloud servers. Executing analytics on IoT data gives insights to make informed and more accurate decisions for IoT applications and machine learning applications. These data, many times, are noisy as the IoTs may reside in noisy environments. The data many times have gaps, invalid readings, and corrupted messages which must be cleaned up before data analysis. This data has to be processed with other relevant data to extract meaningful information. For example, soil humidity data makes more sense with information about whether it rained or not in the agriculture land in IoT-based precision agriculture application. Also running analytics algorithms on large data set requires large computational resources, which are very complex and costly to set up and own. Cloud service on data analytics is the cost-effective way to achieve this.

Major analytics algorithms available on cloud servers are data filters, mathematical transformations (time to frequency and vice versa), and processing with device-specific metadata (data about data, e.g., metadata of Microsoft Word file, the author, when was it created, access rights, where is it stored, etc.) like from which device id, device type, location, timestamp, etc. before storing the processed data on cloud. Data analytics can also extract machine learning information and add intelligence to it.

Data analytics service will be managed by the service provider and can be scaled up as the IoT data grows and as IoT devices are added to the network.

7.8.4.1   Data Reporting and Visualization

The raw and processed data is presented in a friendly format for user to see, hear, or consume, in tables, graphs, or other visual formats. Sometimes, it is converted into regional languages and displayed or spoken. Alexa is one example for such a service.

7.8.4.2   Cloud Application Program Development

The cloud service providers offer a comprehensive set of development kits for using their services on a variety of standard operating systems. It includes cloud development kits (CDK) where the modules are available as plug and play, repository, standard development kits (SDK) on embedded platforms, simulators, fault simulators, settings for different programming language support, etc. An examples is Amazon's Cloud9 which is a development platform for AWS cloud services. Alternative IDEs available are Eclipse, Microsoft Visual Studio, Codenvy, Jupiter Notebook, Particle, etc., depending on programming language preference.

7.8.5   Access to Cloud Servers

Cloud services are accessed by IoT devices by establishing links at the transport layer of standard Ethernet OSI layer stack. The communication happens on *connection-oriented link* using *TCP/IP protocol* for reliable *data transfers* and on *connectionless link* using *UDP protocol* for *voice data transfers*. More about this is covered in later chapters.

7.9   Edge Development Platforms

Edge computing is a part of computing topology in distributed computing in IoT system in which data computing is carried out closer to the edge devices or things. Edge computing brings data storage and the data computing facilities close to the IoT devices so that data collected by the device get processed locally in real time rather than transferred to cloud servers located far away from things. Applications such as virtual and augmented reality, self-driving cars, smart cities, health monitoring, and even building automation systems require fast processing and response and, hence, are the good candidates for edge computing. This offers low latency for responses for the devices.

There are many edge development platforms available from different vendors. AWS Greengrass, Microsoft Azure IoT Edge, and Google Cloud IoT Edge are major software platforms, which support edge computing by enabling cloud computing resources locally. These platforms are software-only platforms like AWS Greengrass, Google Cloud IoT Edge, or combination of hardware and software like Microsoft Azure IoT Edge. The edge functionalities can be implemented on any standard high-performance multi-core systems or laptops or devices. Edge device will get connected to IoT devices using any of the connectivity protocol and uses the Internet to connect to other edge device on edge hub or cloud.

7.9.1 Protocols Supported Depending on Connectivity Technology Chosen

Different protocols are to be supported in software according to the chosen connectivity technology. For long-range communication using LoRaWAN, standard applicable is the LoRaWAN® specification [2], which is a low-power, wide-area (LPWA) networking protocol designed to wirelessly connect battery-operated "things" to the Internet in regional, national, or global networks, targeting key Internet of Things (IoT) requirements such as bi-directional communication, end-to-end security, mobility, and localization services.

Connection to cloud server is facilitated by MQTT or XMPP or CoAP and HTTP protocols. These protocols operate over transport layer TCP/IP of standard Ethernet stack as defined by IEEE 802.3 standard [4]. The data transfers between the IoT device and the cloud happens in JavaScript object notation (JSON) or extensible markup language (XML) format.

More about communication protocols will be dealt in further chapters.

7.9.2 Access to Cloud Services

Depending on the application requirements, level of quality of service, and cost of solution, cloud services can be accessed by subscribing to the preferred service provider. Limited cloud services like limited cloud storage, SQL database management services, IoT device registration, user registration, and data visualization services are freely available for trial which can be explored, and later when the solution has to be scaled up or commercialized, subscription can be availed.

On user subscription by registration, there are many easy-to-use resources available to avail these services. As an IoT system solution case study, the data visualization of environment monitoring parameters temperature/humidity is demonstrated using ThingSpeak cloud service from MathWorks.

7.9.3   IoT Mobile Application Development Platforms

When offering IoT system solution, it is necessary to consider the IoT ecosystem including the mobile application requirements. This is because of the access and afford-ability to smart devices like mobile phones, tablets etc. Most IoT application platforms support mobile app development for data-driven services, predictive analytics, and cloud data processing. Additionally, it acts as an integrated development environment (IDE) that gives the users a useful toolkit for the purpose of IoT application development. These toolkits help to provide end-to-end solutions while assisting in developing, deploying, and managing IoT applications.

Some of the well-known IoT mobile application development platforms are:

- Amazon Web Services
- Microsoft Azure IoT
- Cisco IoT Platform
- Google Cloud IoT
- HP Enterprise Universal IoT Platform
- IBM-Watson IoT Platform
- Predix IoT Platform
- ThingWorx IoT Platform
- Samsung ARTIK
- Qualcomm's IoT Development Kit

Most of the platforms support IoT application development across operating systems, flexibility for scalability, and Industrial IoT scenarios. They also provide optimal solution for application development, management of devices, real-time data analytics, and cloud database management.

References

1. www.bluetooth.com
2. https://lora-alliance.org/
3. White Paper on IOT Firmware for Connected Devices by Thinxstream
4. www.ieee.org

Chapter 8
IoT Security

8.1 Need for IoT Security

As the IoT devices are proliferating in every domain, it is obvious that there is lot of data (both private and sensitive) communication which happens on the Internet and there will be equal scope to misuse the information for various reasons. Adoptability and scalability of the IoT device in any domain depend on how secure it is in preserving the privacy of user information and also access to control the device. Point of access to the network via the internet is seen as weakling and an entry point for an unauthorized access to everything on the network. Wikipedia defines the term cyber-attack as any attempt to expose, alter, disable, destroy, steal, or gain information through unauthorized access to or make unauthorized use of an asset. As more and more activities including critical ones are being done on connected platforms, there is exponential growth in a number of cyber-attacks and cyber-crimes. 78% of US and 93% of Indian organizations have experienced cyber-attacks in year 2020 which is the highest record in recent times [1]. The UK ranked at 5 in cyber-attacks and cyber-crimes [2]. This has enforced these countries to develop IoT security regulations for IoT solution providers to comply with the standards. The EU Network and Information Security Directive (NISD) supervises and monitors IoT compliance to security. Singapore has its own Cybersecurity Act (CSA), and in the USA, the National Institute of Science and Technology (NIST) sets out standards for safe, secure product realization. The main objective of all these standards is to have end-to-end security with authorized access, reliable computation, secure communication, and safe data storage. It is only possible to achieve this compliance if IoT device and M2M communication are made *secure by design*. Typical IoT network is shown in Fig. 8.1.

© The Author(s), under exclusive license to Springer Nature Switzerland AG 2021   123
V. S. Chakravarthi, *Internet of Things and M2M Communication Technologies*,
https://doi.org/10.1007/978-3-030-79272-5_8

Fig. 8.1 Vulnerable IoT network

There are three major challenges of IoT security. They are:

1. Interoperability: There is a great need for interoperability for IoT systems as they are expected to work with many types of devices, subsystems, and systems in the network. Implementing security and maintaining it is a greater challenge.
2. Privacy: In applications like e-healthcare, industrial automation, and defense, data privacy is the major challenge which is make or break property of IoT system.
3. Security: Security is dependent on trust which is a major property and equally challenging as IoT systems involve many different types of users in a variety of use cases.

8.2   Security, Privacy and Interoperability Challenges in IoT Systems

The following sections define the different challenges IoT systems pose in terms of privacy, security, and interoperability.

8.2.1   Challenges of Privacy

The concept of privacy has to be referred to only personal data in the context of IoT. Standard body *oneM2M* guidelines define personal data as "Any information relating to an identified or identifiable natural person ('data subject'); an identifiable natural person is one who can be identified, directly or indirectly, in particular by reference to an identifier such as a name, an identification number, location data, an online identifier or to one or more factors specific to the physical, physiological, genetic, mental, economic, cultural or social identity of that natural person."

The key challenges for privacy in IoT systems can be summarized as follows:

- Profiling and characterizing an IoT user due to the large spectrum of types of people around.
- The understanding one's right on data ownership due to hyper-connectivity hindering transparency of data shared. For example, it is very difficult to know intended user of the mail content on email server.
- The criticality of consent is not understood by everyone to the same extent. Either it is freely given or it is not given.

Therefore, in an IoT environment, it is difficult for someone to exercise control over his/her information.

8.2.2   Challenges of Security

It is not just IoT device security, but the entire system including devices, hub, gateway, and cloud has to be considered to assess the security goals. Thus, IoT devices should verify that it is connected to a known device on the network known to it. This is possible only by device identification and authentication, and the data collected and communicated is trustworthy and correct (e.g., in healthcare IoT system, data sensed by a blood pressure sensor is from a reliable and accurate sensor). Also, information collected is protected for eavesdropping (not corrupted or modified) during transit which is achieved by the method called encryption-decryption.

8.2.3   Challenges of Interoperability

The IoT systems are made of a large number of devices of different types. The system is dynamic and scalable always in which the devices are disassociated and new devices getting connected freely. These devices can be using old technologies and new technologies, and hence backward and forward compatibility is another feature of these networks. The user applications will always be evolving, and newer applications are getting added all the time. The IoT systems have to allow these changes dynamically. This dynamism places new challenges of interoperability at every level.

8.3   Terminology

Trust: Trust in the context of IoT is the reliable communication, storage, accuracy, and data integrity by multiple entities on the network including applications.

Security: The information is sent across entities to the intended recipient from the source sender such that only they can read and interpret without any other device able to read, corrupt, or interfere with it during transit. This is security.

Privacy: In spite of billions of IoT devices access the Internet, the information should belong to only the originators and intended users only and no other devices.

Encryption: Encryption is the process of generating new set of information from original data and the secret key. The secret key has to be available only with the authorized sender and the receiver nodes.

Decryption: Decryption is the reverse process of encryption. It is the process of deriving plain data from the received data and the secret key.

Secret key: Secret key is the random number of fixed length used by encryption and decryption engines in sending and receiving entities in a network for encoding/decoding the data.

Root of trust (RoT): A system entity that provides services, including verification of system, software and data integrity and confidentiality, and data (software and information) integrity attestation between other trusted devices in a system or network [3]. Root of trust is achieved by the device identity and secret key to be used only by participating devices in communication. These should not be able to changed by any other device on the network.

Physical unclonable functions (PUF): This is the device identity uniquely generated using the physical variations in the device fabrication and is not clonable. This is used when the highest level of security is needed.

8.4   Vulnerabilities in IoT Network

Vulnerability of IoT network is weak points in the network through which any unauthorized device (hardware or software) can manipulate the information belonging to others. National Institute of Security and Threat (NIST) defines vulnerability as a weakness in the computational logic (e.g., code) found in software and hardware components that, when exploited, results in a negative impact to confidentiality, integrity, or availability. Mitigation of the vulnerabilities involves design changes, including specification changes or deleting even the supported functionality (e.g., removal of affected protocols or functionality in their entirety). Vulnerabilities are weaklinks through which are unauthorized access of device or resource or data hacks happens for some reason, with intention of causing harm or disturbance or for creating nuisance. IoT network hack can happen from any weak entry points from device to communication protocol or software application. Just to visualize the effect of vulnerabilities of IoT solution, let us consider an example of smart city which is the future of our living. It promises to provide access to civic services, healthcare, education, and public and private services at our fingertips. It is a complex ecosystem of municipal services, public and private entities, people, processes, devices, and city infrastructure that constantly interact with each other. It uses digital technologies with IoT and M2M communication as a core player. Cyber risks impact can be so severe that it can destroy the advantages of smart cities. The consequences could extend beyond just data loss, financial impact, and reputational damage risks—severe enough—to include disruption of crucial city services and infrastructure across a broad range of domains such as healthcare, transportation, law enforcement, power and utilities, and residential services. Such disruptions can potentially lead to loss of life and breakdown of social and economic systems.

Another example of system vulnerability is in autonomous cars. Consider fully autonomous care whose controls (like breaks, engine, accelerators, inner car conditions) can be accessed by anywhere anytime. One can imagine the effects if it is hacked and is in the hands of a criminal. Similarly, one can imagine the impact of network hack by an unauthorized person even in manufacturing or prediction houses. One can refer to more realistic cyber-attacks in reference site of Deloitte [4].

It is absolutely essential to keep it safe and secure by design, installation and during operation. This is needed to be tracked in real time as IoT network is dynamic with devices entering the network and exiting the network is very common. This massive amount of data exchanges, integration between disparate IoT devices, and dynamically changing processes creates new cyber threats, compounded by complexities multiple technologies and multiple ownerships of the entities in such solutions. This requires cybersecurity to be built in in every link in addition to the device-level security as shown in Fig. 8.2.

Owing to the security needs of such networks, the standardizing body oneM2M has defined the method to provide *end-to-end security and group authentication* for the oneM2M complying systems. The candidates in end-to-end security are shown in Fig. 8.3.

The scope includes use cases, threat analyses, high-level architecture, generic requirements, available options, evaluation of options, and detailed procedures for executing end-to-end security and group authentication. This involves multiple functionalities:

- *End-to-End Authentication* means to provide an entity with the ability to validate another entity's identity that was supplied as part of the message before interaction with it. Another entity could be reached with multiple hops in the network.
- *End-to-End Data Confidentiality Protection* provides the ability for an entity to provide protection of the confidentiality of the data in spite till the data reaches destination entity via multiple hops consisting of both protected and unprotected entities. This means that data has to be protected by the entity when it is stored in any device (at rest) and during communication (in transit).
- *End-to-End Data Integrity Protection* provides the ability for an entity to protect data integrity. Data integrity means to retaining the data as it was before transmit-

Fig. 8.2 IoT secure network

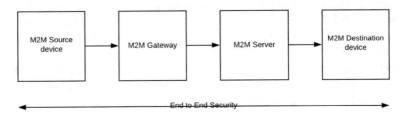

Fig. 8.3 Candidates in end-to-end security

ting without any manipulation while communication till it reaches the authorized
consumer of the data.

- *End-to-End Security* provides the capability to the entities for securing messages
  that can be transmitted via multiple hops before it is received by the authorized
  destination entity. This involves mutually authenticating the entities which wants
  to communicate with each other. It is a method to protect confidentiality and
  integrity of the data to be communicated between the authorized devices.
- *Group Authentication* provides an entity (authenticator) with the ability to validate the identities of all entities which belong to a particular group.
- *Object-Based Security* is a technology that embeds application data within a
  secure object that can be safely handled by untrusted entities without ham-
  pering it.

8.5   Cryptographic Algorithms for IoT Security

Following two mandatory features are essential to make IoT systems secure and reliable:

1. The IoT devices must have unique identities which cannot be compromised. With the device identity, it is possible to identify the sender and receiver without any ambiguity. The identity has to be unforgeable, clonable, and immutable. This is the primary requirement of adopting cryptography methods to secure connection. This will establish the root of trust in the IoT system.
2. IoT device should be able to send the message to the intended recipient device only which can interpret the information in spite of multiple hops via devices both protected and unprotected. This requires additional information called secret key which are held by only sender and receiver which can decrypt the messages.

There are three types of keys which are used in cryptographic algorithms: They are asymmetric, symmetric, and hashing functions.

All these can be used in secure IoT systems.

1. Asymmetric cryptography: This is a type of cryptography which uses two keys: a public key and a private key. The public key is held by all the entities of the network in the group, and the private key is the unique key held by the individual entities. The encryption of the message by the sender is done by the public key, but the intended recipient device with the private key can only decrypt the message. All the devices in the network will know that the device with the matching public key has sent the message but only the device which carried the private key exchanged with the sending device can decrypt the message.

   The device authentication is carried out for identifying the device by generating digital signatures and certificates and for private key exchange.
2. Symmetric cryptography also known as symmetric encryption is the type of cryptography which uses the same secret key for both encryption and decryption of the message during communication between two connected devices. Symmetric keys are typically 128-bit or 256-bit length. During this process, the data is converted to a format that cannot be read or inspected by anyone who does not have the secret key that was used to encrypt it. It is faster than asymmetric cryptography.

   The most common example of symmetric cryptography is advanced encryption standard (AES).
3. Hash functions are used for data integrity checking, ensuring that the received data has not been tampered with during transit. It is a mathematical function which generates fixed length hash values from message to be transmitted.

Level of security on the IoT network is guaranteed by the unique identification of endpoint device and the quality of cryptographic keys generated and exchanged between connected devices. Unique identification of the IoT device guarantees root of trust (ROT) in crypto terminology. Encryption-decryption mechanism using strong keys guarantees the confidentiality of the messages over the communication channel. These have to be managed throughout the life cycle of the IoT devices on the network. This is achieved by the IoT security platform which manages security on the IoT network.

8.6   IoT Security Platforms

There are IoT security platforms which are used to manage the IoT device on the Internet connected to cloud server. Figure 8.4 shows IoT security platform.

The functions of IoT security platform are the following:

1. Provisioning the endpoint IoT devices

    Device provisioning is configuration of the IoT device for the intended function. For example, provisioning a IoT device for temperature-humidity involves configuring the settings and configurations with necessary reset values and initialization tasks for its control/configuration registers, thresholds for right classifications and memory sizes for storage of captured data. Before the device actually starts its intended function, the firmware should be able to generate the unique identity of the device and the required cryptographic keys required to store the measured data securely in the device. It is also expected to generate the necessary certificates for the authentication for onboarding it onto the Internet.

2. Onboarding those devices to a server so that they can access applications

    The authentication certification is sent to the server for verification, and on confirmation, the device is connected to the server on the Internet. The secure connection is established and the device data is sent to the server.

3. Managing the devices throughout their life cycle

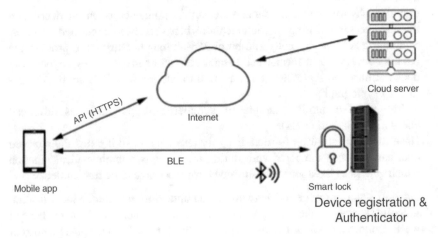

Fig. 8.4 IoT security platform

On the secure link, the device can access any services it has subscribed on the network, to throughout the life cycle of the device. The other functions of management of IoT devices include monitoring the functional integrity and ensuring that they have not been compromised by a cyber-attack, firmware signing, and encryption to enable and secure updates over the air (OTA).

4. Disassociation of the endpoint IoT device when needed

Key and certificate renewal, which are needed for reconfiguration when it is rebooted in the network. Disassociating and Resetting and re-onboarding the device on the network with new initialization when it is compromised has to be correctly managed.

The IoT security platforms can be local network with the edge device or sophisticated on demand cloud service platform equipped with AI and machine learning algorithms. Figure 8.5 shows conceptual cloud-based security platform.

In cloud based security platform, the metadata regarding IoT devices, traffic patterns, manufacturer details, and logging in pattern are collected via firewall and logged into the main network server. IoT security cloud collects these metadata corresponding to the IoT device, applies machine learning and AI algorithms, and use them to identify most of the devices on the network. This learning happens every time the IoT device is active on the network. Any abnormal pattern observed

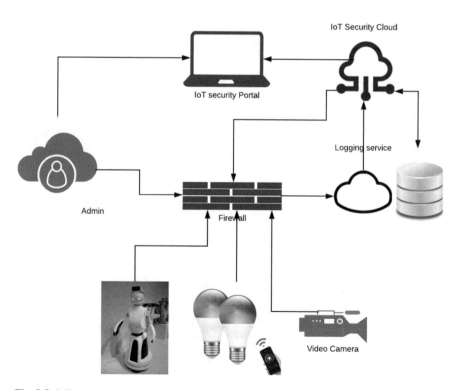

Fig. 8.5 IoT security cloud service

suddenly is informed to the administrator by message alerts to avoid any unauthorized access. The security portal is updated with the device information which is accessed by the administrator. Some platforms also supports *IoT security tomography*. Network tomography refers to real-time learning of distributed vulnerabilities and security aspects by monitoring the activities on the network.

8.7   Layer-Wise Security of IP Stack for IoT

IoT device consists of the functions corresponding to layer functions and corresponding protocol implemented as hardware and software or combination. Typical IoT solution in a layer architecture as a modified OSI layer along with different protocol landscape is shown in Fig. 8.6.

IoT security is having a unique, immutable, and unforgeable identity for each device and being able to send data securely between devices using a combination of cryptographic functions. The two main features which are needed for ensuring security in communication are (1) ensuring strong root of trust (ROT) which is the device identity and (2) the cryptographic keys which are not compromised by unintended entities during communication.

Since IoT device connects to the internet through the network layer, security protocols of transport layer along with network layer security will be the most appropriate for IoT ensuring end-to-end security in network.

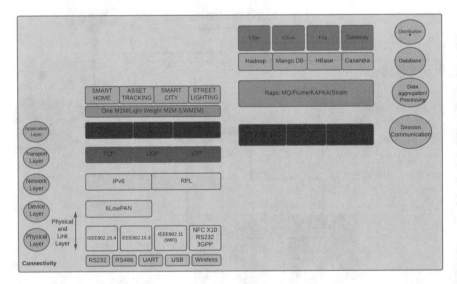

Fig. 8.6 Layer architecture and protocol landscape of IoT system

8.7.1   Transport Layer Security (TLS)

In Transport Layer Security (TLS) protocol, the messages are encrypted between web application and servers. This gets executed when the web browser is loading a website, emails, messaging, and VoIP. This procedure for IoT applications is defined in Datagram Transport Layer Security (DTLS) and TLS1.3 protocol, which are implementable on many constrained IoT devices. DTLS is used for UDP and TLS for TCP protocols. DTLS and TLS provide security between the client application and server by operating between the application layer and transport layer. It ensures encryption, authentication, and data integrity of data communication. HTTP based software applications with TLS implementation is referred to as HTTPs. TLS connection is established by sequence of handshake messages between client application and server. During these handshake transactions, the client and server negotiate the version of TLS both the devices support, generate **session key**, and **authenticate** the server. Authentication mechanism uses certificate which is obtained by certificate authority (CA) which is generally third party. This uses asymmetric cryptography and can use multiple hops between them.

8.7.2   IPsec for IoT

IP security (IPsec) protocol protects the security of IPv6 headers by implementation at transport layer protocol in network layers. This is used between the sensor device and the Internet hosts unlike multi-hop TLS security. This can be used as end-to-end security option for wireless sensor networks like 6LoWPAN or IEEE802.15.4 technologies. Also, the protocol used message compression techniques for multi-hop in WSNs.

8.7.3   Cryptography Keys

End-to-end security is achieved by the effective implementation of cryptography algorithms with strong cryptographic keys. These keys range from 128 bits to 256 bits in symmetric cryptography and 2048 bits in asymmetric cryptography depending on algorithm and layer at which it is implemented. Generation of these keys is an essential feature of any crypto algorithms. The cryptographic keys are the unique random numbers which are generated and associated with the device identity. There are advanced techniques like quantum-driven energy-based random generation used to preserve the identity of the device and preserve necessary root of trust. As a number of devices get connected to the internet, to ensure that device identity is not compromised, techniques based on uniqueness in hardware is used for key generation. The random key generation algorithms also use device fabrication parameters, like oxide thickness, real time on-chip temperatures as seed values, which are truly unique to the device. This technique is called physical unclonable functions (PUF). It is necessary to ensure that these keys are generated and stored in the device in the secured storage and exchanged with the intended peer device over the secured channel.

8.8   IoT Security Tomography and Layered Attacker Model

The assessment of vulnerability of IoT network is called IoT security tomography. This is carried out by using the layered attacker's model for IoT network. This provides sections and subsections of network which are vulnerable to attacks by monitoring complex network activities. The three-layer attacker's model is shown in Fig. 8.7.

8.8.1   Layer 1: Physical or Device Layer

Physical or Device layer is also called **perception layer**. IoT devices are connected with many types of sensors and actuators. In this layer, functions processing all sensitive information which are collected by the sensors or controls actual physical parameters surrounding it are executed. The threats in this layer are the device layer attacks. A major threat at this layer is **eavesdropping**, which is accessing private communications like emails, SMSs, and video calls and using them to their advantage. Another hazardous attack in the perception layer is **node captures** where the attacker can seize access to sensors or manipulate the sensor data or take control of the device. The third type of attack is the **fake and malicious node** which is an

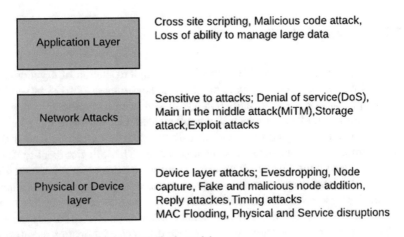

Fig. 8.7 Three-layer architecture attacker's model

attack in which an attacker adds a node to the system and inputs fake data. It aims to stop transmitting real information. A node added by an attacker consumes precious energy of real nodes and potentially controls them in order to destroy the network. *Reply attack* is an attack in which the attacker uses authentic data from the sender and replies back to get the root of trust. The attacker sometimes even disrupts the services. The attack can be through personal area network (PAN) technologies like Wi-Fi, Bluetooth, NFC, and PLCs.

The attacks at the perception layer can be prevented by *key-based encryption and sensor data protection methods.*

8.8.2   Layer 2: Network Layer Attack

The network layer is the bridge between the perception layer and the application layer in the three-layer architecture model of IoT. Since it is the entry point of IoT to the Internet, it is very sensitive to attacks. Common security threats in this layer are *denial-of-service* (DoS) attack, where the attacker prevents the authentic device to access the services it has entitled to on the network *man-in-the-middle attack* (MiTM), where the attacker intercepts and accesses sensitive information from the ongoing communication and misuses it; *storage* attacks, where the attacker manipulates the stored data of the devices in the network; and *exploit* attack, which is an illegal attack gaining control of the network.

Network layer attacks can be avoided by proper *identity authentication* and encryption mechanisms, and by configuring the network switches with DHCP or Spanning Tree Protocol (STP), and implementing additional checkpoints like (1) ARP inspection, (2) disabling unused ports, (3) enforcing effective security on VLANs (virtual LANs) to prevent VLAN hopping, and (4) provisions for message access security (MAS), root key data store, devices, and data authentication in LWM2M functions for device gateway to the Internet.

8.8.3   Layer 3: Application Layer Attack

Application layer attacks are *injection attacks* called *cross-site scripting* which is scripting false information; *malicious code*, which is a code in any part of the software intended to cause undesired effects and damage to the system; and *loss of ability to process* large data which can make the applications slow.

Application layer attacks can be blocked by using antivirus software, securing cloud computing, and authentication and privacy protection methods.

8.9   Best Practices in Monitoring IoT Security

To maintain IoT security, it is necessary to follow some of the following best practices for large networks:

- Daily check security alerts that are received in the emails or SMS notifications and respond as appropriately based on their severity and urgency.
- Review system alerts in the IoT security portal and the firewalls page to check that firewalls are connected to IoT security. If a firewall is disconnected, IoT security stops analyzing log data, and no new device detections and identifications occur. Serious events that increase risk to the IoT devices in the network could be missed. Also watch unusually large log files generated from firewall. This can be an indication of anomalous activity.
- Scan the services which are newly discovered and confirm that they are authorized to use the network. Unauthorized devices pose a threat if they did not undergo an onboarding process.
- Keep a track on the network activity of high-value devices on the network.
- Check that the firewall regularly receives IP address-to-device mappings from IoT security to ensure no devices are missing from policy enforcement.
- Whenever there is a network expansion, look for new network segments and update with more firewalls to provide added segments.

References

1. https://www.broadbandsearch.net/blog/alarming-cybercrime-statistics
2. https://mobilenewscwp.co.uk/Unified-Comms/article/number-cyber-attacks-uk-businesses-surge-2020
3. www.wikipedia.com
4. https://www2.deloitte.com/us/en/insights/focus/smart-city/making-smart-cities-cyber-secure.html

Chapter 9
IoT Application Technologies for Constrained Devices

9.1 Introduction to IoT Application Layer and External Interface

Application software is the interface for the IoT device to the user. In short, it is called application. It interacts with the user in the environment to display the status of the device about what is being processed or processed parameters of user's interest for which the device is intended or control the parameters as the case may be in the IoT device. In IoT user interaction with the system, take place through the applications. The intent of the application layer differs depending on the the functionality of the device in the IoT network solution. For example, in gateway device, the application layer is used to configure the device. On the cloud software, the application layer serves as user interface. This application software corresponds to the application layer of OSI architecture. There are many types of applications needed to interact with the IoT network. Different application layer functions needed in IoT systems are the following:

IoT device applications to interface and interact with the user, physical environment to collect, capture, monitor the parameters and to control the selected parameters. Also, from application layer functions, it gets configuration parameters and provides processed information to the user via this interface.

Web application software connects the device to the Internet and run on gateway device or servers. These applications are large and complex software, targeted on unconstrained environment with complex processors and require large memory.

Application software, which collect data in the IoT devices run on the constrained environment with small microprocessor and limited memory.

IoT things route the data collected through these applications to the Internet. The two types of environments on which applications are developed are constrained RESTful environment (CoRE) and unconstrained environment. This application generally runs on gateway devices. These applications need special constrained environment protocols such as CoAP or Light weight machine to machine (LWM2M) protocol.

© The Author(s), under exclusive license to Springer Nature Switzerland AG 2021
V. S. Chakravarthi, *Internet of Things and M2M Communication Technologies*,
https://doi.org/10.1007/978-3-030-79272-5_9

9.1.1   CoAP Application Protocol

CoAP application protocol is customized application protocols for IoT devices. This protocol is defined by IETF, which is simple, lightweight, and operate with the *request/response mode* of the communication. It is an enhancement of HTTP protocol targetted for resource optimization, which is run on constrained low-power devices. It is based on the successful REST model, in which *resources* are available under a URL and clients can access those resources using the *GET, PUT, POST, OBSERVE,* and *DELETE* commands. The centralized resources are data set stored in centralized memory, storage memory or certain functions for running analytics on dataset, etc. The protocol is suitable for large IoT networks. CoAP protocol also support *publish-subscribe* protocol with *extended GET* command. CoAP protocol is used only when the transport layer is *UDP on IPv6* on lossy low-power PAN networks like 6LoWPAN or LW-PAN. CoAP is used to create and manage the devices on network, publish-subscribe to service on the network, and indicate the status (power on-off) when requested by the peer devices. Most of the application development platforms support CoAP protocol. Some of the known development platforms are Zetta, Arduino, OpenRemote, Node-RED, Flutter, ThingsBoard, Kinoma, SiteWhere, Kaa IoT platform, and M2MLab's Mainspring. Typical use case scenario of CoAP application is shown in Fig. 9.1.

IoT network *myoffNW* shown in Fig. 9.1 is for temperature monitoring of two office desk environments. The two desks are **desk 1** and ***desk 2*** where IoT devices

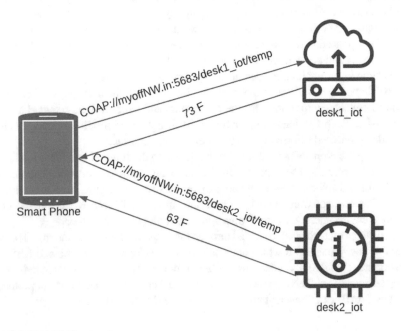

Fig. 9.1 CoAP IoT network

desk1_iot and *desk2_iot* are connected, respectively. The devices are accessible by a smartphone where the application is running. The devices also run applications which monitor the temperatures around the desks. The communication protocol used in this case is CoAP protocol as the temperature monitoring devices **desk1_iot** and **desk2_iot** are constrained devices. The commands used in the example are shown on the arrows shown in the figure. The CoAP devices always use port with number **5683**. In the request command *COAP://myoffNW.in:5683/desk1_iot/temp* sent, *myoffNW* is the IoT network domain, and *desk1_iot* is the sensor name at **desk 1** of the office. Similarly, when the request is sent to *desk 2*, the same port number **5683** is used, on network *myoffNW*, and the sensor name is *desk2_iot.* The return values of temperatures at *desk 1* and *desk 2* devices are *73 F* and *63* F, respectively.

CoAP devices send status codes corresponding to different status of the device as listed in Table 9.1.

Table 9.1 CoAP status codes

CoAP status code	Description
2.01	Created
2.02	Deleted
2.03	Valid
2.04	Changed
2.05	Content
2.31	Continue
4.00	Bad request
4.03	Forbidden
4.01	Unauthorized
4.04	Not found
4.05	Method not allowed
4.06	Not acceptable
4.08	Request entity incomplete
4.12	Precondition failed
4.13	Request entity too large
4.15	Unsupported content format
5.00	Internal server error
5.01	Not implemented
5.02	Bad gateway
5.03	Service unavailable
5.04	Gateway timeout
5.05	Proxying not supported

9.1.2   Message Queueing Telemetry Transfer (MQTT) Application Protocol

IoT solutions communications involve transmitting information beyond user data. They include:

- Real-time events generated in the IoT devices.
- Captured and processed information from the sensor devices for publishing. Sharing published information with other devices in the network on subscription.
- Polling for real-time events on a specific sensor device of interest.
- Short messages without much overhead and they are targeted to constrained devices.
- Communication over unreliable channel for noisy devices and less reliable noisy networks.

The existing HTTP based application layer protocol could not manage the listed requirements of IoT communications efficiently. This prompted IBM to develop MQTT protocol which is message queueing telemetry transfer (MQTT) protocol.

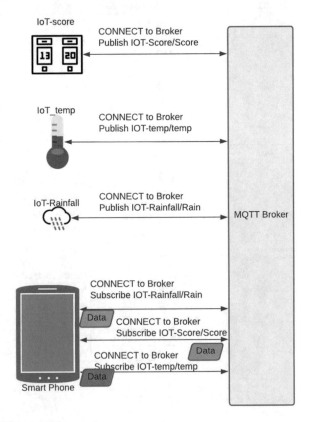

Fig. 9.2 MQTT-based IoT network

This is a *lightweight simple reliable protocol for IoT networks*. The MQTT protocol runs on the TCP protocol of the transport layer. It uses *subscribe-publish* mode of communication. The devices *publish the services*, and the other devices seeking these services *subscribe* to them through the *broker* to access them.

Typical use case scenario using MQTT protocol for IoT is shown in Fig. 9.2. In the figure shown, there are three IoT devices which publish the data captured by them to the broker. IOT-Score sends score information the container (database storage) called IOT-Score. IOT-temp keeps publishing the temperature to IOT-temp container in the broker. IOT-Rainfall publishes information regarding the rainfall to IOT-Rainfall container in the MQTT broker. Three applications in smartphone are

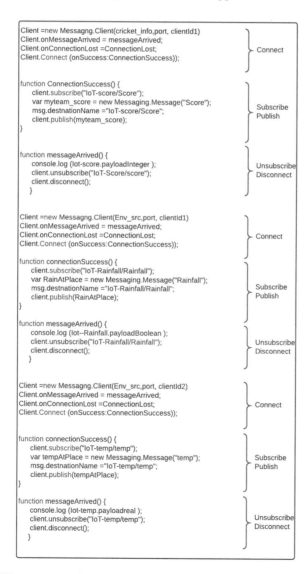

Fig. 9.3 MQTT message transfers for the application

```
Client =new Messagng.Client(cricket_info,port, SmartPhone)
Client.onMessageArrived = messageArrived;
Client.onConnectionLost =ConnectionLost;                          } Connect
Client.Connect (onSuccess:ConnectionSuccess));


function ConnectionSuccess() {
    client.subscribe("IoT-score/Score");
    msg.myteam_score ="IoT-score/Score"
    }

function connectionSuccess() {
    client.subscribe("IoT-Rainfall/Rainfall");            } Subscribe
    msg.RainAtmyPlace ="IoT-Rainfall/Rainfall";
    }

function connectionSuccess() {
    client.subscribe("IoT-temp/temp");
    msg.tempAtmyPlace ="IoT-temp/temp";
    }


function messageArrived() {
    console.log (IoT-Rainfall.payloadBoolean );
    console.log (IoT-temp.payloadreal);
    console.log (IoT-score.payloadInteger);          } Unsubscribe
    client.unsubscribe("IoT-Rainfall/Rainfall");       Disconnect
    client.disconnect();
    }
```

Fig. 9.4 MQTT transfers for a smartphone device

rainfall collector, score collector, and temperature collector. Smartphone user through these MQTT applications can subscribe to one or more information from the containers in the broker.

The sequence of the publish/subscribe transfers in the application scenario is shown in Fig. 9.3.

MQTT message transfers for smartphone are shown in Fig. 9.4.

MQTT supports three levels of security which will be QoS 0, QoS 1, and QoS 2 as shown in Fig. 9.5.

MQTT devices support two types of wildcard subscriptions * and ? for single-level or multi-level services.

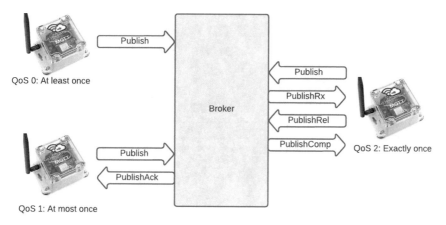

Fig. 9.5 MQTT security

9.2 Extensible Messaging and Presence Protocol (XMPP)

Instant communication of messages (messaging applications like WhatsApp, GTalk) use protocols which are Extensible Messaging and Presence Protocol (XMPP) and WebSocket. XMPP is an open-source, decentralized network (anyone can set up and run the XMPP server) protocol, used for M2M communication. It is based on extensible markup language (XML), which allows the protocol to be extended for a wide variety of purposes. XMPP is most used for instant messaging (IM) like WhatsApp and Telegram applications where the messages are instantly transferred. It is also called *Jabber* Protocol. The main advantage of this protocol is extendibility, and hence it can be used for many proprietary applications which used XMPP. XMPP applications are flexible and support content syndication, collaboration tools, file sharing, gaming, remote systems monitoring, web services, lightweight middleware, and cloud computing, beyond IM. Some of the key XMPP technologies are:

- Core—information about the core XMPP technologies for XML streaming.
- Jingle—SIP-compatible multimedia signaling for voice, video, file transfer, and other applications.
- Multi-user chat—flexible, multi-party communication.
- Pub/sub—uses publish/subscribe flow for alerts and notifications for data syndication, rich presence, and more.
- Bosh—it is "Bidirectional-streams Over Synchronous HTTP," an HTTP binding for XMPP (and other) traffic.

IETF has defined standard for core XML protocol. Typical use case is shown in Fig. 9.6. The XML message in XMPP protocol is called **stream.** As shown in the figure, the devices with unique identity transfer message streams in XML.

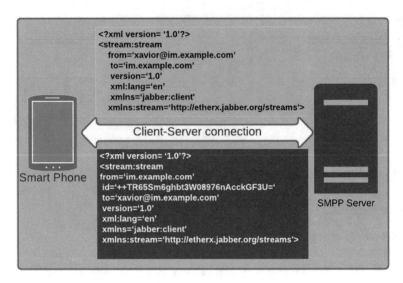

Fig. 9.6 XMPP client-server connection

The basic unit in XMPP is called **stanzas**. There are three types of stanzas:

Message stanza: It is used to send messages. The format of message stanza in XML is shown below:

> *<message>*
>
>> *from='vani@myserver.org'*
>>
>> *to='juliet@sense.com' id='msg_1'>*
>>
>> *<body>If you read this message then you have received it.</body*
>>
>> *<data xmlns='http://jabber.org/protocol/ibb' sid='ft_session_2' seq='84'>*
>>
>>> *Jkfsjfklsjfklasjflkajflkajfkljfalsjflkasjfajf'ja'jafjkfkjkshafjkhsfkjahf*
>>>
>>> *fksafklflkjklajfjflkajflkjdfjdklajfkljnflkjflkjkjfkljfkljfkjdjfkdjkdjfkj*
>>
>> *</data>*
>
> *</message>*

Presence stanza: It is used to know the status of the peer device.

```
|-------------------|
| <stream>          |
|-------------- -----|
| <presence>        |
|  <show/>          |
| </presence >      |
|-------------------|
| <message to='foo'>|
|  <body/>          |
| </message>        |
```

• Info-query stanza: It is used to *get* some information from the server or to *set* or apply some settings on server or client device. The format is shown below:

```
|------------------|
| <stream>         |
|------------------|
|------------------|
| <iq to='bar'>    |
|   <query/>       |
| </iq>            |
|------------------|
| ...              |
|------------------|
| </stream>        |
|------------------|
```

Detailed study material for XML language is not the scope of this book but may be referred at [2].

9.2.1   WebSocket

WebSocket is another full duplex messaging protocol which is a centralized server architecture. This is used for large file transactions as shown in Fig. 9.7. It offers true concurrency and hence offers best performance compared to XMPP. This is also used for M2M communication using client-server setups. In this, all the devices are connected to the server, and the server acts as the mediator for communication. The disadvantage is the security as it is built on transport protocol of IP stack. This is used for real-time audio-video streaming, which needs high performance.

Typical code snippets of WebSocket are shown in Fig. 9.8. For detailed information on WebSocket, one can refer to [4].

The advantages of the WebSocket protocol are its human-readable code style, high performance, and suitability for proprietary applications. The only concern is the security.

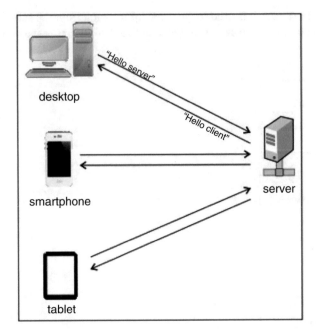

Fig. 9.7 WebSocket configuration

```
// Create a new WebSocket.
var socket = new WebSocket ('ws://echo.websocket.org');

socket.onopen = function(event)
{   console.log("Connection established");
// Display user friendly messages for the successful establishment of connection
    var.label = document.getElementById("status");
    label.innerHTML = "Connection established";
}

socket.onmessage = function(event)                    socket.onmessage = function(event)
{                                                      {   if(typeOf event.data === String )
  if(event.data instanceof ArrayBuffer )                {
  {   var buffer = event.data;                             console.log("Received data string");
      console.log("Received arraybuffer");             }
  }                                                     }
}
```

Fig. 9.8 WebSocket code snippets

9.2.2 Simple Object Access Protocol (SOAP)

SOAP is another XML-based protocol which uses HTML to access web services. There are applications developed in Java, .NET, and PHP. SOAP was developed as an intermediate language so that applications built using different programming languages could communicate easily among each other. It is a platform-independent lightweight protocol which can operate across programming languages as it works on HTML. SOAP message consists of header and message body constituting XML page. Typical code snippet of SOAP message is shown in Fig. 9.9. For more details, one can refer to [5].

9.3   Representational State Transfer (REST)

Representational state transfer (REST) protocol is a simpler alternative to SOAP. In this protocol, the constraints are applied to data elements, components, connectors, and objects so that they are not transferred to the server making it really lightweight by reducing the complexity and improving performance. This enables scalability of devices on network and server which is needed for IoT networks. User interface and user state are of no relevance for the server, and hence REST allows to block them sending them to the server. Also, REST constraints support client-server architecture and layered architectures. Client-server configuration improves scalability by simplifying client instances on server. Layered configuration permits connectivity through any intermediate layers like firewalls, gateways, proxies, or intermediate servers. REST makes intermediate layers process messages with constraints such that their interaction is self-descriptive.

```
<xsd:complexType>
    <xsd:sequence>
        <xsd:element name="Tutorial Name" type="string"/>
        <xsd:element name="Tutorial Description" type="string"/>
    </xsd:sequence>
</xsd:complexType>


<soap:Body>
    <GetTutorialInfo>
        <TutorialName>Web Services</TutorialName>
        <TutorialDescription>All about web services</TutorialDescription>
    </GetTutorialInfo>
</soap:Body>
```

Fig. 9.9 SOAP message structure

9.3.1   RESTful and HTTP APIs

When all the transactions between the machines are constrained with the REST protocol, it is called *RESTful API*. Standard *HTTP APIs* are *GET, PUT, POST*, and *DELETE*. HTTP-based *RESTful APIs* use same commands of HTTP, viz., *GET, PUT, POST*, and *DELETE* with the identifiers: *URI/URLs*.

9.3.1.1   Application Development Platforms

Following are some of the development platforms used to develop IoT applications which support all the protocols discussed above:

- Zetta: API-based application development platform based on Node.js for IoT and web application development. It can make any device as API developed for real-time and data-intensive devices. It can be used to develop applications which can run on PC, server, and any medium complexity hardware platforms.
- Arduino: It is the appropriate combination of hardware-software which is an open-source development environment which supports application development.
- OpenRemote: It has introduced a new open-source IoT platform to create professional energy management, crowd management, or more generic asset management applications.
- Node-RED: It is a visual tool to developing IoT device organization. It is built on Node.js used for wiring together hardware devices, APIs, and online services.
- Flutter: Easily programmable hw-sw solution based on Arduino flutter with wireless connectivity up to half a mile without router.
- M2MLabs Mainspring: It is an application framework. It is a Java-based framework used to develop machine-to-machine (M2M) applications such as remote control, fleet administration, or smart terminal. Its facilities include flexible design of devices, device structure, connection between machines and applications, validation and normalization of data, long-term data repository, and data retrieval functions.
- ThingsBoard: It is an IoT application development framework used for data collection, processing, visualization, and device management. It upholds all standard IoT protocols like CoAP, MQTT, and HTTP as quickly as cloud and on-premise deployments. It builds workflows based on design life cycle events and REST API events.

There are many other application development frameworks from Marvell Semiconductor (Kinoma), Kaa, SiteWhere, and many others. One thing to note is that all these platforms support onboarding of devices using multiple protocols discussed in this chapter and scalability and other performance depend on the server and communication resources which these use in hardware.

References

1. www.ietf.org
2. https://xmpp.org/rfcs/rfc3920.html
3. www.xml.org
4. http://websocket.org/
5. https://www.w3.org/TR/soap/

Chapter 10
M2M Communication and Technologies

10.1 M2M Communication

Direct communication between two machines or two devices without human involvement is called machine-to-machine or M2M communication. M2M communication can be wired or wireless. This communication between devices will be for monitoring, controlling, or communicating partially processed information for further processing in a remote machine in an integrated system solution. No single technology or solution is ideally suited for a variety of applications, market situations, and available spectrum for IoT communication. There are many wired and wireless technologies, which enable M2M communication. Combination of one or more of these technologies are adopted to achieve communication across machines and devices. (Devices are considered as machines here.) Different wired technologies available for M2M communications are:

- Serial connections
- Power line connections

Wired communication technologies are used to achieve more reliable communication when the machines are very close (in a few meters of distance) to each other and require external power supplies. With the advancements in wireless technologies, offering same or better reliability and mobility, wired technologies are not preferred nowadays.

Different wireless technologies used for M2M communication are the following:

- Cellular:

  - LTE
  - Cellular IoT NBWAN

- Satellite communication

© The Author(s), under exclusive license to Springer Nature Switzerland AG 2021
V. S. Chakravarthi, *Internet of Things and M2M Communication Technologies*,
https://doi.org/10.1007/978-3-030-79272-5_10

- Short-range technologies:
  - Bluetooth
  - Wi-Fi
  - Zigbee
- LPWAN:
  - LoRa
  - Sigfox
- GPS/GNSS and positioning

10.1.1   Cellular Technology

The cellular technology used for M2M/IoT is Long-Term Evolution (LTE) also known as fourth-generation mobile technology (4G), which all the service providers offer on their networks. LTE has two diverging branches of technology:

- LTE Cat 1 to Cat 18: Cat 1 to Cat 18 represent user equipment (UE) categories used to define the specifications of LTE devices, which offer high-speed data throughput, ideal for primary internet and high-speed data communication from one machine to the other, e.g., multimedia communication and office network over cellular network.
- LTE Cat 0, Cat M1, and NB-IoT: LPWAN LTE (low-power, wide-area) delivers a new class of cellular connectivity for low-speed, wide-coverage IoT /M2M applications using very low power to communicate the information between two machines. Cat M0 and Cat M1 are used for such large IoT-based networks for M2M communication in small multiple bursts of data communications. This uses very low current for such communication and hence has low power consumption.

The latest of LTE is 5G technology, which provides two types of services. The core 5G networks offer speeds faster than the current 4G LTE. The LTE Cat M1 and NB-IoT compliant to 5G networks provide very long network life, low data rate, and low power and support burst transfers, which are suitable for IoT M2M communication.

10.1.2   Satellite Communication

M2M communication when there is no terrestrial coverage is achieved by employing satellite communication. Such solutions typically use dual-mode communication with LTE as primary communication and, when the machine goes out of terrestrial coverage area, switch to satellite mode for data communication over wider geographic coverage. The examples are mining, monitoring, transport tracking, etc.

10.1.3 Short-Range Technologies

There are many short-range technologies available for M2M communication. Among them, Bluetooth, Wi-Fi, and Zigbee are the most used technologies for M2M communication. They constitute WPAN networks. They offer over-the-air (OTA) communications with good reliability within a short range and are self-configuring networks. These technologies are suitable for in-premise machine-to-machine communication. However, the Bluetooth 4.0 called BT-LE offers long-range communication at very low data rate.

10.1.4 LPWAN Technology

LPWAN is a wireless technology with a wide-area network with low data rates of the order of hundreds of kilobits per second (Kbps) used to interconnect low-power battery-operated devices suitable for IoTs. Range covered by this technology is 2–1000 km. LPWAN technology is a group of technologies using RF spectrums in licensed and unlicensed bands. LoRa, Sigfox, NB-IoT, and LTE-M are few of the well-known LPWAN technologies competing among them for IoT solutions.

10.1.5 GPS/GNSS and Positioning Technology

The Global Positioning System (GPS) and Global Navigation Satellite Systems (GNSS) are systems using satellite communication for accurately positioning and ranging systems which are used in IoT systems.

Depending on the application, any of the above technologies is used for M2M communication solutions. Table 10.1 gives the technical comparison of some of the M2M communication technologies.

10.2 Standards and Protocols

To enable communication between a device and any of the networked devices, there is a need for interoperability. Interoperability is guaranteed by complying to standard defined by the European Telecommunications Standard Institute (ETSI) [2], through **oneM2M**, an initiative [3]. **oneM2M** is a global standards initiative that covers requirements, architecture, API specifications, security solutions, and interoperability for M2M and IoT technologies. M2M device is IoT device with communication capabilities. An M2M device hosts one or more M2M applications

Table 10.1 Comparison of specification of wireless M2M communication technologies

Sl. no.	M2M communication technologies	Frequency of operation	Data rate MBPS	Range	Power usage	Cost
1	Cellular 2G/3G	Cellular	10 Mbps	Hundreds of kilometers	High	High
2	Bluetooth	2.4 GHz	1,2,3 Mbps	~100 m	Low	Low
3	LPWAN 802.15.4	SubGHz, 2.4 GHz	40,250 kbps	>350 km^2	Low	Low
4	LPWAN: LoRa	SubGHz	<50 kbps	1–5 km	Low	Medium
5	LPWAN: Cellular LTE CAT0/1	Cellular	1–10 Mbps	Several km	Medium	High
6	LPWAN: NB-IoT	Cellular	0.1–1 Mbps	Several km	Medium	High
7	LPWAN: Sigfox	SubGHz	<1 kbps	Several km	Low	Medium
8	LPWAN: Weightless SiG	SubGHz	0.1–24 Mbps	Several km	Low	Low
9	Wi-Fi	2.4 GHz, 5 GHz, 60 GHz	0.1–54 Mbps	<100 m	Medium	Low, medium
10	Zigbee	2.4GHz	250 kbps	~100 m	Low	Medium

or other computing applications and can contain implementations of common service entity (CSE) functionalities.

The general architectural framework of any M2M device is shown in Fig. 10.1.

As per the architecture proposed by oneM2M, any M2M device has two domains: **device domain** and the **network domain**.

According to the oneM2M standard, communication between two M2M entities takes place as shown in the functional architecture shown in Fig. 10.2.

As in Fig. 10.2, the M2M communication happens between two internet protocol (IP)-capable devices. Originator device comprises functionalities corresponding to application layer services, which in turn uses underlying IP network capabilities using transport layer protocols on UDP and TCP to communicate the data to the peer M2M device. The application layer of M2M device accesses the transport layer protocol of Internet stack using either hypertext terminal program (HTTP) for web application, Constrained Application Protocol (CoAP) (for lightweight M2M (LWM2M) device), or Message Queueing Telemetry Transport (MQTT) protocol.

Layer-wise functional architecture of M2M device with IP and non-IP connectivity is shown in Fig. 10.3.

The Open Connectivity Foundation (OCF) defines specifications for data models for different profiles in different vertical domains of M2M device.

As one can see in Fig. 10.3, the network connectivity in non-IP device will be done using **layer 2 (L2)** connectivity which provides the functionalities required for establishing physical and data link layer connections (e.g., Wi-Fi or Bluetooth

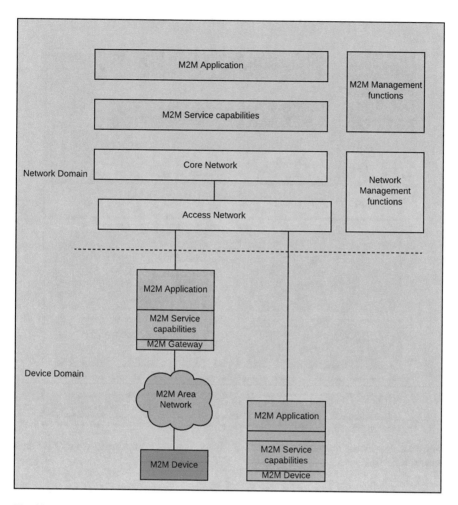

Fig. 10.1   M2M architecture framework

connection). **Networking** provides functionalities required for devices to exchange data among themselves over the network (Internet). **Transport** provides end-to-end flow transport with specific quality-of-service (QoS) constraints. **Framework** provides the core functionalities like identity assignment, resource model, device management, etc. as shown. The functional communication block uses the set of requests and responses that are the content of the messages between two devices. **Vertical Domain Profile** provides market segment-specific functionalities, e.g., functions for the smart home market segment.

When two devices communicate with each other adopting client-server roles, each functional block in a device interacts with its counterpart in the peer device as shown in Fig. 10.4.

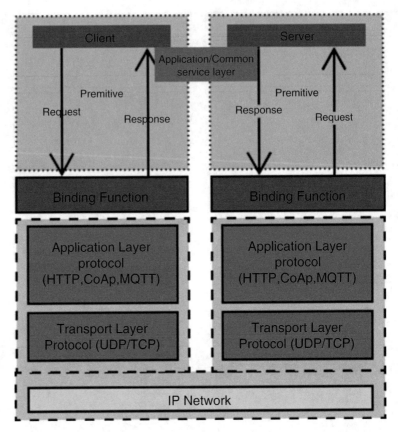

Fig. 10.2 Functional architecture of IP-based M2M communication. (Courtesy: oneM2M draft standard, ETSI)

Functions of each layer in the *Framework* shown in Fig. 10.4 are as follows:

1. *Identification and addressing*
   Identification and addressing layer defines the procedure to indentify and address the device.
2. *Discovery*
   Discovery defines the process for discovering the devices, their resources and their capabilities in forming the network.
3. *Resource model*
   Resource model specifies the capability for representation of entities in terms of resources and defines mechanisms for manipulating the resources. The resource model function is necessary for identifying the interoperable device around it.
4. *Create, Read, Update, and Delete and Notify (CRUDN)*
   CRUDN provides a generic scheme for the interactions between a client and server configuration of the device.

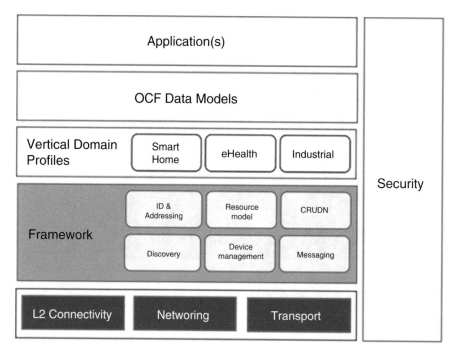

Fig. 10.3 Layered functional architecture of M2M device with IP and non-IP connectivity. (Courtesy: oneM2M standard)

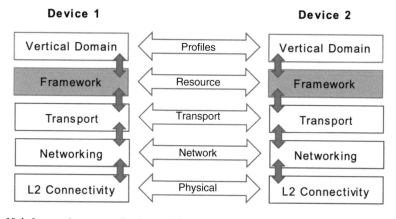

Fig. 10.4 Layer-wise communication model

5. *Messaging*

   Messaging provides specific message protocols for Representational State Transfer (RESTful) operation, i.e., CRUDN. For example, CoAP is a primary messaging protocol.

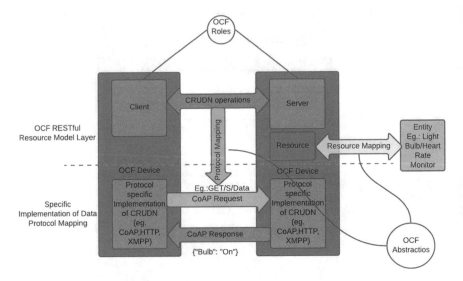

Fig. 10.5 Conceptual architecture of M2M device

6. *Device management*

Device management specifies the method to manage the capabilities of a device, including initial setup, monitoring, and diagnostics.

7. *Security*

Device security includes authentication, authorization, and access control mechanisms required for secure access to entities.

Conceptual architecture of different devices and the roles they take during communicating between them are shown in Fig. 10.5.

10.3   Applications Using M2M Communication

As M2M communication is an important technology along with IoT for realizing automation and connectivity, there are a large number of applications one can adopt in any domains. The major domains that see potential role of M2M communication are the following:

- Smart Cities: The modern cities need to evolve and become a structured, interconnected ecosystem where all civic services (energy, mobility, buildings, water management, lighting, waste management, environment, etc.) are logistically connected and are working together to support comfortable living. By using the IoT and M2M technologies, the cities are expected to achieve this transition maintaining security and privacy, reducing negative environmental impact in a reliable, future-proof, and scalable manner.
- Smart Living (e.g., smart house): For smart living, it is expected that all machines are connected, automated, and easily controlled by people. This could be

achieved by interconnecting and automating machines, equipments, and appliances boosting the comfort of the people. This vertical domain also includes healthcare management using connected things.

- Smart Farming and Food Security: The application of M2M communication and IoT technologies to the overall farming value chain will improve its optimization and, as a consequence, food security in general. M2M communication in this is used for data gathering, processing, and analytics as well as orchestrated automation technologies supported by IoT.
- Smart Wearables: The integration of intelligent systems to bring new functionalities into clothes, fabrics, patches, aids, watches, and other body-mounted devices will provide new opportunities for M2M applications. Fundamental technologies such as nano-electronics, organic electronics, sensing, actuating, localization, and communication are enablers of these applications which will enhance personal comfort.
- Smart Mobility (smart transport/smart vehicles/connected cars): Self-driving, connected vehicles, multi-modal transport systems, and "intelligent" transportation infrastructure deploy M2M communication as one of the technologies in providing comfort mobility infrastructure.
- Smart Environment (smart water management): IoT with M2M communication will be a key building block to solutions for applications such as environmental monitoring and control that will use sensors to assist in environmental protection by monitoring air and water quality/quantity monitoring along water infrastructure (including water resource management), atmospheric or soil conditions, and noise pollution.
- Smart Manufacturing: M2M communication technology will massively benefit the manufacturing industry, by adding more and more artificial intelligence to advanced connected objects providing sensing, measurement, control, power/energy/raw material management, and communication capabilities.

A few of the examples from different domains are discussed here to give perspective of how M2M communication is addressed in these applications.

Figure 10.6 shows the communication between smartphone and thermostat for temperature monitoring application.

The message sequence chart during M2M communication of this application is shown in Fig. 10.7. In this example, devices have to support multiple application layer protocols. The smartphone (client) sends a message in HTTP format to the gateway (intermediary) which converts it to a CoAP message and sends to the server (thermostat) and receives CoAP response from it, which is sent back to the smartphone as an HTTP response. It should be noted that this is communication between a device supporting HTTP and another device with CoAP protocols through intermediatory gateway which supports both.

Another example of tiger monitoring on smartphone application using, positioning engine, and the tiger with the wireless tag is demonstrated in Fig. 10.8.

For the real-time spotting and tracking of tigers, the message sequence chart is shown in Fig. 10.9. The positioning device is mounted at fixed locations where tigers are probably spotted. The tigers are tagged with Low energy bluetooth (BLE)

Fig. 10.6 Smartphone sending a message to a thermostat

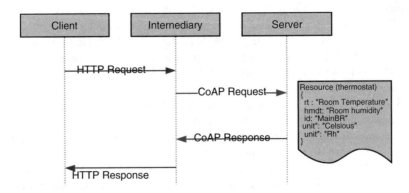

Fig. 10.7 Entities taking client (smartphone), intermediatory (gateway), and server (thermostat) roles during the communication

technology enabled tags. When tigers with BLE tags come in the vicinity of the positioning engine, they respond to the inquiry messages which are broadcasted periodically from the positioning devices. When such response is received, an alert message is sent to the mobile application of the concerned authority. The IoT solution in this case uses Bluetooth 4.0 (low-energy location tracking profile) for the real-time location tracking of tigers. BLE technology is best suited for this application as tigers get spotted rarely. Long-range low-bit rate communication as.

Interactions between the logical entities called roles in the applications are shown in Fig. 10.10.

Room light control or playing a preferred song in home automation/infotainment system using intelligent song players like Alexa use probably the sequence of messages as shown in Fig. 10.11.

Interactions between the logical entities in different devices playing different roles in the application is shown in Fig. 10.12. Corresponding message sequence chart is shown in Fig. 10.13.

Alexa like artificial intelligence (AI)-based home assistant device is registered using mobile application which subscribes to the services available on the cloud server. This uses Wi-Fi on mobile application, AI assistant, and Wi-Fi modem. Thereafter, any request from the user is communicated to the AI assistant through Bluetooth or Wi-Fi network, and then the AI assistant decodes the request and contacts the cloud service for the response from the cloud services, and based on the response, action is taken. If the user request is to play a song, the AI assistant

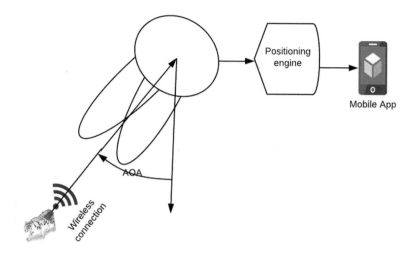

Fig. 10.8 Real-time location search system

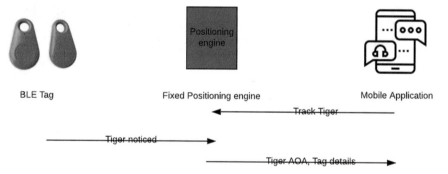

Fig. 10.9 Message sequence chart

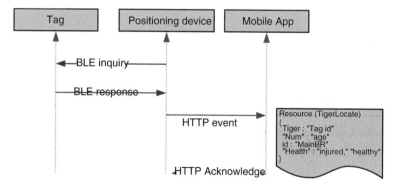

Fig. 10.10 Interaction between roles in livestock tracking

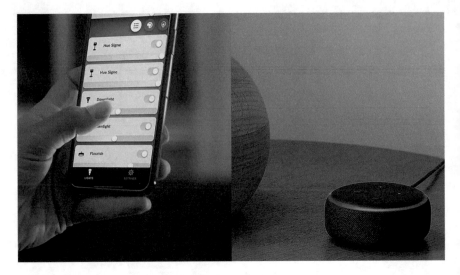

Fig. 10.11 AI-assisted controlled home light control system

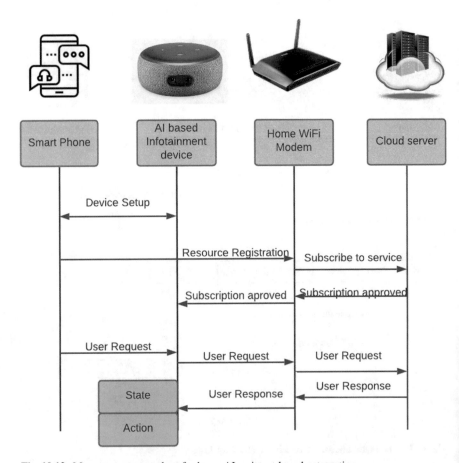

Fig. 10.12 Message sequence chart for home AI assistant-based automation

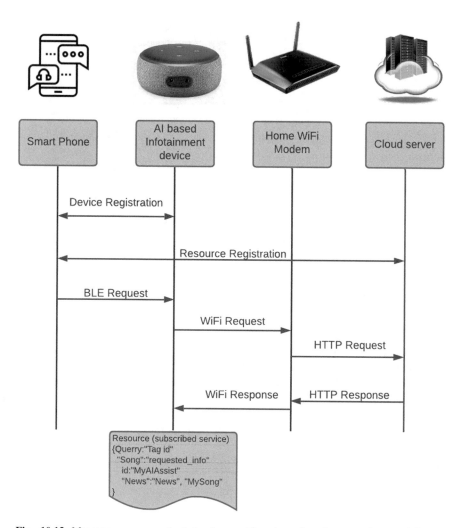

Fig. 10.13 Message sequence chart for home AI assistant-based automation or infotainment system

decodes the request and communicates to the cloud, and the response with necessary controls is received from the cloud service, and the song is played on the device.

IoT based resources like staff/patient/equipment tracking system in a hospital is shown in Fig. 10.14. To track the hospital resources, it is necessary for them to have BLE-enabled tags as wearable or attached to the equipments being tracked. The position locator is mounted in places where it has a good range of radio frequency (RF) signal coverage. Whenever a patient or a hospital staff comes in the vicinity of the position locator, it gets connected to BLE tag and a message is sent to the locator by the wearable device or equipment tag. The BLE-enabled positioning device will periodically send inquiry messages for which the tag or wearable responds. The

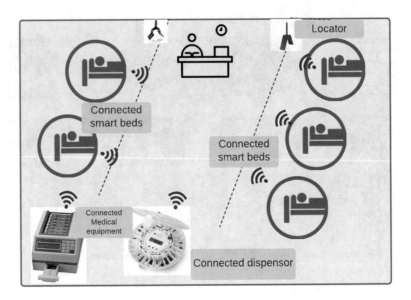

Fig. 10.14 Hospital/patient/equipment tracking system

message received by the positioning device is processed and it is further transmitted to the hospital admin or laptop of doctor/mobile application. The message transfer happens using HTTP protocol or Representational State Transfer (REST) protocol. The data collected regarding equipment will further be analyzed to find out where they belong to and traced further for its effective use. The M2M communication in the hospital positioning system is used by the hospital for providing quality service. In addition, this can improve the security of hospital assets, reduce the spread of infection, reduce costs, and increase the time clinical staff spent with patients.

Interactions between the logical entities in the applications are shown in Fig. 10.15.

Another interesting multi-technology M2M application is monitoring fatigue of workers in mining which typically operates 24 × 7 as shown in Fig. 10.16. This requires all workers to wear a communicating tag or wearable supporting multiple technologies, through which they can communicate. When the worker crosses a cellular network area, the message is communicated over satellite communication to reach the control server.

In all applications discussed, the M2M communication happens over multiple network topologies using technologies either through the Internet or with non-Internet communication technologies like Bluetooth, Wi-Fi, satellite communication, or WAN as shown in Fig. 10.17.

Figure 10.18 shows different technologies used in applications in different segments of M2M communication.

M2M connectivity through the Internet is through the IPv6 layer. A more detailed discussion on IP stack and IPv6 layer will be done in a subsequent chapters. By adopting right combination of communication technologies, a broad spectrum of

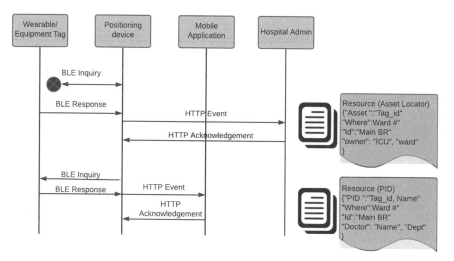

Fig. 10.15   Interaction between different machines in hospital positioning system

Fig. 10.16   Fatigue monitoring in mining (Source: PRECO Electronics)

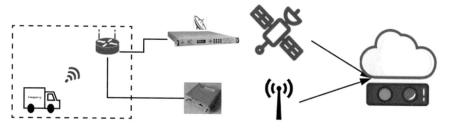

Fig. 10.17   System framework for asset or resource tracking using multiple communication technologies

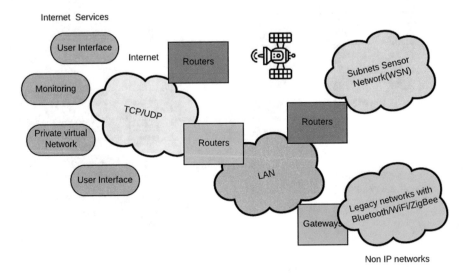

Fig. 10.18 Different technologies through which M2M communication is achieved

useful IoT based network can be deployed. These products can range from battery-powered nodes transmitting 100 bytes per day with the ability to last tens of years on a coin cell battery to mains-powered nodes able to maintain megabit video streams. The IpV6 layer functionality of ethernet stack is highly adaptive to connect to technologies like Bluetooth, Wi-Fi, 802.15.4, and LPWAN and hence the most common way to connect any device or machine for communication. The router in the figure 10.18 is an IPv6 router or edge router or, very rarely, satellite router only for non-terrestrial communication-based M2M connectivity. Machines limited in processing power, memory, non-volatile storage, or transmission capacity require special IP adaptation layers (6LoWPAN) and/or dedicated routing protocol for low-power and lossy networks (RPL). Examples are devices transmitting over low-power physical layer like IEEE 802.14.5, Bluetooth Low Energy, and near-field communication (NFC). In such cases, communication between devices across IP and non-IP networks requires devices with dual-mode protocol converters between IPv6 and non-IPv6 protocols.

To understand any M2M communication therefore, it is important to understand the protocols used to communicate between the device domain and the IPv6 and non-IPv6 network domains.

References

1. www.openconnectivity.org
2. www.etsi.org/standards
3. www.onem2m.org

Chapter 11
M2M Communication Protocols

11.1 Internet Protocol Stack and M2M Communication

M2M connectivity needed in most IoT-based solutions in different applications connects to the Internet using Internet protocol (IP) at the transport layer. The data rate supported in M2M communication will range between 100 bytes per day in battery-powered IoTs and multiple megabytes per second in high speed real-time video streaming kind of applications. Figure 11.1 shows the device communication framework in the application accessing TCP/IP suite to connect to the peer machine on internet through the routers.

11.2 Internet Protocol (IP) Stack

Standard communication protocols are defined with reference to the seven-layer OSI protocol stack recommended by ISO-specified Open Systems Interconnection (OSI) shown in Fig. 11.2. The internet protocol (IP) stack in OSI model is shown in Fig. 11.3.

As shown in Fig. 11.3, communication happens in two types of configurations depending on the applications. They are **unreliable connectionless** and **reliable connection oriented**.

Unreliable connectionless configuration supports transmissions without bothering to check if the transmission is successful or whether the receiver has received the transmitted messages. **Reliable connection-oriented configuration** makes sure the receiver has received the transmitted messages. In Reliable connection-oriented configuration, communication carry overheads like retransmissions, acknowledgement from the receiver, etc. to ensure that the data received by the receiver is correct. In applications, where the overhead of maintaining the reliable transmission is

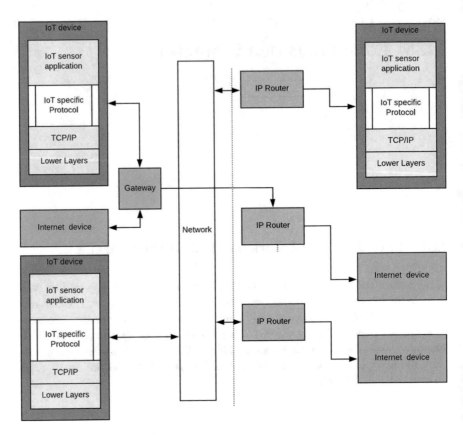

Fig. 11.1 M2M connectivity framework

unacceptable, unreliable connectionless configuration is adopted. The IP layer is the
first virtual protocol abstraction layer in TCP/IP stack. Virtual protocol abstraction
layer connects the transport layer resources to the network layer. The IP layer does
not provide any functionality for error correction of datagrams that are lost in the
channel, received multiple times, or received not in proper sequence. If no such
errors occur in the physical layer, the IP protocol guarantees that the transmission is
successful. Hence, the IP layer is also called **unreliable connectionless** communication. Typically, the functions of error handling, duplicate rejection, and sequence
numbering and recovery are managed by the layers below network layers, viz., data
link and physical layers. Figure 11.4 shows basic IP datagram (packets) fields and
their description.

In the IP datagram, the first five (5) thirty-two (32 bits) fields are called IP header
which are shown in light gray in the figure.

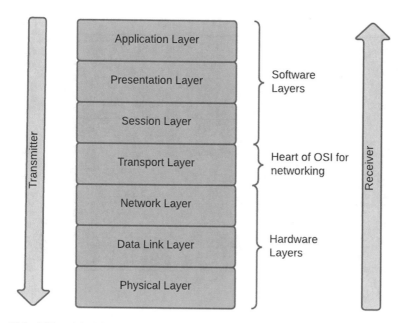

Fig. 11.2   OSI model of Ethernet stack

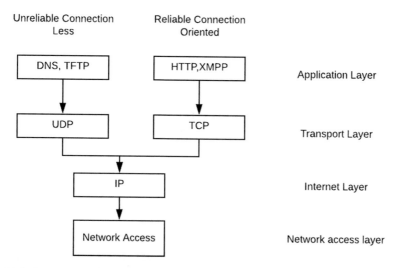

Fig. 11.3   IP stack

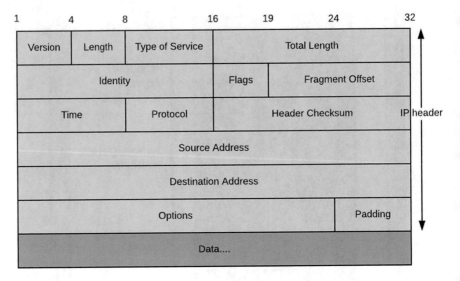

Fig. 11.4 IP datagram

11.2.1   Version

There are two versions of IP datagrams defined, IPv4 and IPv6. The first three bits in the frame header specify the version of the datagram.

11.2.2   Length

It is the length of the IP header. Without any **OPTIONS**, this value is 5. It determines whether the datagram has OPTIONS and PADDINGs or not. Padding means extra bits to make the header 32 bits.

11.2.3   Type of Service

Datagrams can be given 8 priorities as depicted by **TYPE OF SERVICE** which is three-bit field indicating the priorities 0–7 of the data carried in the datagram. This is used to give priorities of data transmission during scheduling the data transfers. This is to allow *out-of-band* data to be routed faster than normal data. This finds relevance as few of the control frames are communicated as IP frames and they need to be transmitted on priority to normal user data. Also, the **TYPE OF SERVICE** field allows a classification of the datagrams as per the service requiring short delay time, high reliability, or high throughput. This is possible only when the data communication happens on multiple routes in the network that get connected to the destination.

11.2.4   Identity, Flags, and Fragment Offset

These fields in the IP header are used to communicate to the receiver if a datagram is fragmented or not. The actual length of an IP datagram is independent of the length of the actual frames being transferred on the network. This is called maximum *transmission unit (MTU) of the network*. If a datagram is longer than the MTU, then it is divided into smaller frames called fragments having the same headers. The IDENTITY flag is used to identify segments belonging to the same datagram, and the FRAGMENT OFFSET is the relative position of the fragment within the original datagram. The frames are fragmented till they are communicated to the receiver so that any missing fragment or repeated fragments are corrected and reassembled at the receiver.

The layers are not hidden below the IP layer in spite of the fragmentation functionality. The reason is that the MTU can vary from 128 or less to several thousands of bytes dependent of the physical network (Ethernet has a MTU of 1500 bytes). Depending on the channel quality and the performance required, the datagrams are fragmented so that a number of fragments are reduced.

11.2.5   Time

TIME field is the remaining *time to live (TTL)* for a datagram when it travels on the Internet. The routing standard specifies that at most 15 hops are allowed before it reaches the destination device.

11.2.6   Source IP Address and Destination IP Address

Both the source and destination address are indicated in the datagram header so that the IP layer of the recipient can send an acknowledgement back to the IP layer of the transmitting device. Note that the IP datagram can accommodate one source IP address and a destination IP address. The Internet layer passes the IP address of the **next hop** address to the network layer below for transmitting the datagram to the next hop device which is the intermediate device. This IP address is bound to a physical address, and a new frame is formed with this address. The rest of the original frame is then encapsulated in the new frame before it is sent over to the communication channel.

11.3   IPv6 and IoT

As shown in the IP stack, there are two types of IP protocols IPv4 and IPv6. The difference is the increased device addressing capability. The IPv4 and IPv6 datagram headers are shown in Fig. 11.5. General Internet datagrams can be of IPv4 or IPv6 types on the Internet. For networks with IoT devices, IPv6 is used.

As it can be seen in the IPv6 header, the address space of source and destination devices is made 128 bits to accommodate a large number of IoT devices getting connected to the Internet. It will serve to address the demand of IoT devices of the future. It is expected that there will be 50 billion connected devices on the Internet by 2025.

11.3.1   User Datagram Protocol (UDP)

The User Datagram Protocol (UDP) is a protocol built on the top of the IP. The basic unit of data is a *user datagram*, and the UDP protocol provides the unreliable, connectionless service similar to IP datagram. The main difference between IP and UDP is that the UDP protocol is an *end-to-end protocol*. That is, it contains enough information to ensure/check that the transfer of user datagram from the transmitting device to the receiving device is complete. The format of a user datagram is illustrated in Fig. 11.6.

The LENGTH field is the length of the user datagram including the header. The minimum value of LENGTH is 8 bytes. The SOURCE PORT and DESTINATION PORT are the connection numbers assigned in the network between the two IP addresses. A network port is an integer number.

When calculating the UDP CHECKSUM, the UDP protocol appends a 12-byte pseudo header consisting of the SOURCE IP ADDRESS, the DESTINATION IP ADDRESS, and some additional fields. When a device receives a UDP datagram, it

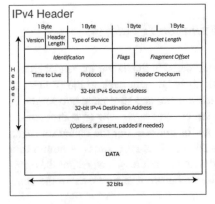 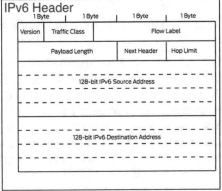

Fig. 11.5 IPv4 and IPv6 headers

Fig. 11.6 UDP extension
of the header

1	16	32

Source Port	Destination Port
Length	UDP Checksum

takes the UDP header and creates a new pseudo header using its own IP address as the DESTINATION IP ADDRESS, the SOURCE IP ADDRESS, and other fields extracted from the IP datagram. Then it calculates a checksum, and if it equals the UDP checksum extracted from the datagram, then the datagram is accepted successfully.

As shown in Fig. 11.3, the UDP protocol is used as the basic protocol in client-server application protocols such as in a simple File Transfer Protocol (FTP) like **Trivial File Transfer Protocol** (TFTP) and domain name system (DNS) where the overhead of making a reliable, connection-oriented transmission is considerable and not desired.

11.3.2 Transmission Control Protocol (TCP)

The **Transmission Control Protocol (TCP)** provides a full duplex, reliable, connection-oriented service to the application layer on IP as shown in Fig. 11.3.

The TCP protocol is designed to transfer a large set of user data from the application layer reliably to the destination device. It establishes a full duplex virtual link (with three-way handshake of start and stop links) between the two devices, so that both can transmit data set on the Internet without explicitly specifying the destination once the connection is established. A modified TCP protocol called **Transactional Transmission Control Protocol (T/TCP)** is defined to suit M2M communication in client-server mode for continuous data transfers in a reliable way.

11.3.3 TCP Segment Format

Basic data unit of TCP protocol is called **TCP segment**. Different fields in TCP segment are presented in Fig. 11.7.

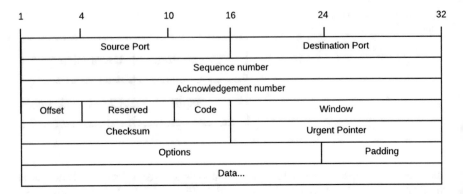

Fig. 11.7 TCP frame format

11.3.3.1   Source Port and Destination Port

The TCP protocol also uses a pseudo header consisting of port numbers instead of the source IP address and the destination IP address. The IP header carries the source and destination addresses.

11.3.3.2   Code

The **CODE** field defines the content of the TCP segment to indicate the end of frame (EOF).

11.3.3.3   Options

The TCP protocol uses the OPTIONS field to exchange information like *maximum segment size accepted* between the TCP layers on the two communicating devices. The flags are as follows:

- **URG** Urgent pointer field is valid.
- **ACK** Acknowledgement field is valid.
- **PSH** This segment requests a push.
- **RST** Reset the connection.
- **SYN** Synchronize sequence numbers.
- **FIN** Sender has reached end of its byte stream.

For more details on the fields and their functions, Internet standard document [13] can be referred.

11.3.3.4   Offset

The offset indicates the relative distance in bits of the user data within the segment. This field has to be used if **OPTIONS** field is used.

11.3.3.5   Urgent Pointer

This field is used to indicate that the control information such as escape codes, etc. are located in the user data. This will be used to process this information at the destination device immediately when it receives the segment.

11.3.4   Reliable Transmission in TCP Link

Practically, it is seen that not all packets transmitted may not reach the IP layer at the destination device due to different reasons like network congestion, noisy channel, and gateway failures. TCP protocol provides reliable communication by following methods to correct errors in packets during transit, deleting duplicate packets received, recovering lost packets, and rearranging out-of-sequence fragments of packets. This is achieved by assigning a SEQUENCE NUMBER to each fragment transmitted and receiving a *positive acknowledgment (ACK)* from the receiving device. If the ACK is not received within a pre-allotted time called timeout, the data is retransmitted. At the receiver, the sequence numbers are used to correctly order segments that may be received out of order and to eliminate duplicates. Packet error is corrected by adding a checksum to each segment transmitted, checking it at the receiver, and discarding damaged segments. The principle is illustrated in Fig. 11.8.

As shown in Fig. 11.8, **device A** transmits a frame 1 to **device B** which is acknowledged by the device B, but the second packet shown in red gets lost before it reaches its destination. As soon as device A transmits a packet, it starts an ack timer which will be reset on receiving the positive acknowledgement from the destination device B. If acknowledgement is not received and the timer expires, timeout is generated indicating the failure of acknowledgment. The device A retransmits frame 2, restarts the timer, and awaits acknowledgement. A limit on maximum number of retransmissions is defined depending on the network load which is experimental. The time between the transmission of a packet and the reception of the acknowledgement packet is called round-trip time (RTT). This means that the *round-trip time (RTT)* can vary from segment to segment in the network. RTT is calculated by recording successive values of packet arrival times and computing recursive mean value with an exponential window to decrease the importance of old values. TCP link needs handshake on every frame transmission. If the acknowledgement to the transmitted packet from the receiver is not received for a number of times beyond configured maximum retry limit, it is considered as the virtual link failure and re-establishes the link is necessary. Some TCP links also support block acknowledgement which is sending single acknowledgment packet for block of received frames from the sender, to reduce the overhead of these handshake mechanisms in communication.

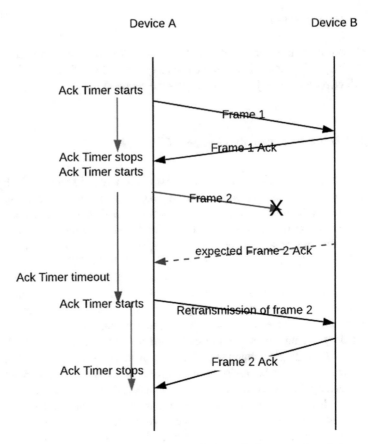

Fig. 11.8 Message sequence chart for transmission of frames, acknowledgement, and retransmission

11.3.5   Transactional Transmission Control Protocol (T/TCP)

The TCP protocol is a highly symmetric protocol in that both devices transmit and receive data simultaneously. However, not all applications are symmetrical by nature. A typical example is a client-server protocol which is an asymmetric modified TCP protocol called **Transactional Transmission Control Protocol (T/TCP)**. This is achieved by stateless (no session information is retained by the transmitter or receiver) request-response-based transfers. It guarantees reliable communication, minimum overhead of link establishment at the network layer, and low-latency responses.

The T/TCP protocol is backward compatible with TCP. However, one of the features of the T/TCP protocol is that it does not require connection start and stop link protocol.

11.3.6   Internet Protocol (IPv4 and IPv6)

Though IPv4 and IPv6 are IP layers on conventional IP stack, IPv6 is proposed to be the IP layer for IoT network on the Internet as it provides large addressing space to connect a large number of IoT devices. IPv6 supports 128 bits addressing compared to 64 bits of addressing in IPv4.

11.4   Application Protocols

11.4.1   HTTP Application Protocol

Access to devices using the World Wide Web (www) browser on the Internet involves access to resources as a large set of documents (wide-area hypermedia information) in HTML format. Basic framework of HTTP communication is in client-server configuration of device or service on the web is as shown in Fig. 11.9. The devices and their resources are uniquely addressed and share resources between them by sharing them as documents on the web.

11.4.1.1   The Client

The client in the Fig. 11.9 is the device which requests to transfer the message to the connected device.

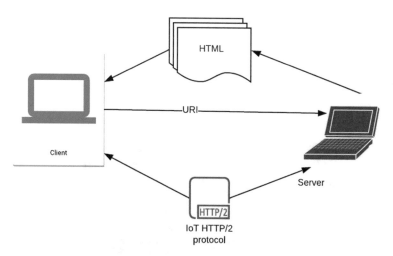

Fig. 11.9 Client-server communication using HTTP protocol using HTML

11.4.1.2   Uniform Resource Identifier (URI)

Uniform Resource Identifiers (URIs) are used to address a data object, location, or any resource in the World Wide Web (www) model. The term "URI" is defined by the IETF. The URI addressing scheme is shown in Fig. 11.10.

The URI addressing space covers the following.

Uniform Resource Locator (URL)

In practice, the URLs consist of the current set of Internet protocols supported by the World Wide Web, i.e., typical URL starts with www, followed by a directory path, a file name, and possibly a search directive in HTTP, FTP, Gopher, WAIS, etc.

Uniform Resource Name (URN)

However, the ultimate goal for URIs is to be a persistent naming scheme independent of the mean of access, i.e., the protocol used, and of the physical structure of resources on the specific host.

Uniform Resource Citation (URC)

This is meta-information about a URI. It consists of pairs of attribute/value which can contain information on the author, publisher, etc. The URC are currently not used.

Server: The device whose URI is requested. It is the device which responds to the request. It can be a service or any resource in the device with URI on the network.

Hypertext Transfer Protocol (HTTP): HTTP protocol is widely used to access web services on the Internet. HTTP is used in many ways. One of those is using HTTP as RESTful API 1256 (HTTP-REST-API). This is a client-server presentation layer for application protocol based on a stateless connection. It is built on a client-server model where the client initiates a request and the server sends a response.

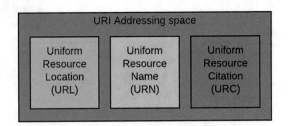

Fig. 11.10   URI address space

The basic format of the HTTP protocol has a set of HTTP headers followed by a message body containing a data object in 7-bit or 8-bit accepted by the client. The client specifies the data format it can handle in the request.

The HTTP protocol provides two methods for the client to transfer a data object to the server, even though the client is not guaranteed that the request will be fulfilled—even on a successful return code. The action can at all times be cancelled by the responsible functional profile of the remote server.

HTTP-REST-API will provide CRUD (Create/Read/Update/Delete) operational primitives naturally with its method (e.g., POST/GET/PUT/DELETE) and well-defined status codes. HTTP-REST-API usually supports extensible markup language (XML) [3] or JavaScript object notation (JSON) for passing API parameters. REST-API can handle various data formats using content-negotiation feature which is part of HTTP specification. It provides easy-to-understand, scalable, secure APIs for web service in distributed computing environments.

11.4.2 Hypertext Markup Language

HTML is the user interface to create information on the World Wide Web. But it is not necessary to store the data in HTML format in the devices. The HTML file can be generated on the fly upon a client request.

The data stack that gets communicated to the other device on the Internet gets transformed to different formats before they leave the physical layer in a device. Different data formats in different layers in the IP stack are shown in Fig. 11.11. The added header fields to the application data by each layer in the data stack are either consumed by the next layer or modified to improve the reliability and data integrity required communication.

The application data in typical HTTP format is converted to TCP/UDP format in the TCP layer and then converted to IP format and then to media access control (MAC) protocol data unit (MPDU) in the media access control (MAC) layer and physical protocol data unit (PPDU) in the physical layer.

The application layer and the physical layer protocols change from technology to technology. Application layer protocols for different platforms are HTTP, MQTT, and CoAP. While HTTP is briefly covered in the above section, MQTT and CoAP protocols will be covered in subsequent sections of this chapter.

The physical layer protocols change depending on the physical medium of communication. Physical medium can wireless or wireline. Most used wireless technologies are Bluetooth, WLAN, Zigbee, cellular or LoRa, and LowPAN. Different standard professional bodies [1, 2] are working on defining new standards or redefining or adding newer features to the existing technologies to suit the needs of IoT and M2M communication. Some of them are the following:

- CoRE: Constrained RESTful Environments
- 6LO: IPv6 over Networks of Resource-constrained Nodes
- 6TSCH: IPv6 over the TSCH mode of IEEE 802.15.4e [i.22]

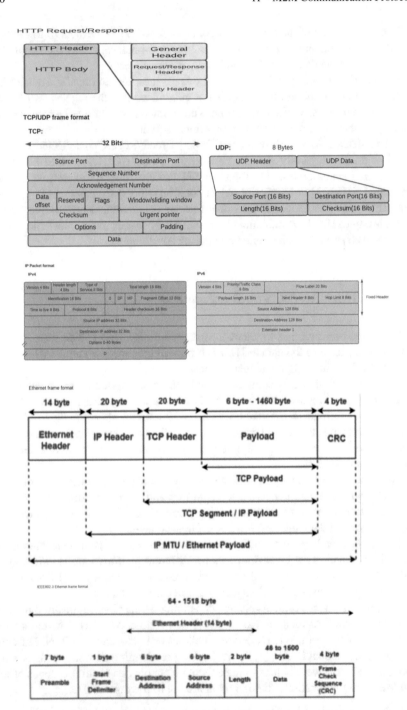

Fig. 11.11 Data stack across the layers

- LPWAN: IPv6 over Low-Power Wide-Area Networks
- RoLL: Routing Over Low-power and Lossy networks
- HOMENET: Home Networking
- ACE: Authentication and Authorization for Constrained Environments
- IPWAVE: IP Wireless Access in Vehicular Environments
- DICE: DTLS In Constrained Environments

11.5   Constrained Application Protocols (CoAP)

The **Constrained Application Protocol (CoAP)** is a specialized web transfer protocol defined for constrained devices on constrained (e.g., low-power, lossy) networks. The constrained devices are low-resource devices which often work on 8-bit microcontrollers with small ROM and RAM storages. The constrained networks are characterized by high packet error rates during communication and low data throughput of the order of tens of kilobits per second. The protocol is designed for machine-to-machine (M2M) applications like smart energy and building automation. CoAP provides a request/response interaction model between the devices, supports built-in or self-discovery of services and resources around them, and supports web access. CoAP devices are designed to easily interface with HTTP for integration with the web by meeting specialized requirements such as multicast support, very low overhead, and simplicity for constrained environments. The CoAP protocol is defined using a subset of REST which is common with HTTP but optimized for M2M applications. The additional features needed for interoperability with HTTP are expected to run on constrained device architectures of low-end microprocessor and memory specifications. CoAP protocol can be made to work on UDP and is intended to be used for M2M/IoT applications. CoAP is also well suited for wireless sensor network nodes. It can be appended with DTLS to provide application layer security. CoAP client-server transmission is shown in Fig. 11.12.

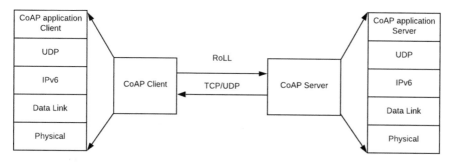

Fig. 11.12 M2M communication on web using TCP/UDP transport layer in constrained environment using RoLL protocol

The key features of CoAP are:

- It is a RESTful protocol.
- It defines four request response communication means similar to HTTP. They are Get, Put, Post, and Delete.
- It supports three types of response code: 2.xx (success), 4.xx (client error), and 5.xx (server error).
- It supports four different message types: Confirmable, Non-confirmable, Acknowledgement, and Reset (Nack).
- It supports both synchronous message exchange and asynchronous message exchange via observe/notifications. It is a web access protocol fulfilling M2M requirements.
- UDP binding with optional reliability supporting unicast and multicast requests. Confirmable and Acknowledgement/Reset messages to provide optional reliability when required.
- It has low header overhead and parsing complexity.
- It supports URI and content-type support.
- It is a stateless HTTP mapping, allowing proxies to be built providing access to CoAP resources.
- It supports security through **Datagram Transport Layer Security (DTLS)**. A wide variety of key management mechanisms may be used for this purpose.
- It supports a resource directory mechanism where IoT/M2M devices (i.e., CoAP servers) can register/update their list of resources. It stores only URIs to resources stored on servers. If a device is in sleep mode and not able to communicate with the network, it can be discovered via this resource directory.
- Group-based communication using (unreliable) IP multicast data transfers.
- Stateful observation intends to reduce overhead in the network due to multiple re-registration requests from CoAP client to CoAP server when the server (i.e., IoT/M2M device) is not in a position to accept additional clients.
- It supports mechanism to transport CoAP over SMS for cellular networks.
- It also supports representation of links in JSON format in an unconstrained environment.

11.5.1 CoAP Protocol Stack

The interaction model of CoAP is similar to the client/server model of HTTP. Typical CoAP implementation supports both client and server roles. A CoAP request is sent by a client to request an action (using a method code) on a resource (identified by a URI) on a server. The server then sends a response with a response code; this response may include a resource representation.

Unlike HTTP, CoAP deals with these interchanges asynchronously over UDP. CoAP defines four types of messages: Confirmable, Non-confirmable, Acknowledgement, Reset; method codes and response codes included in some of these messages make them carry requests or responses. Requests can be carried in

Confirmable and Non- confirmable messages, and responses can be carried in piggy-backed Acknowledgement messages. One could think of CoAP logically as a two-layer approach, a CoAP messaging layer used to deal with UDP and the asynchronous nature of the interactions and the request/response interactions using method and response codes. The CoAP protocol stack is shown in Fig. 11.13.

11.5.2   CoAP Data Format

CoAP allows to explicitly indicate payload of the content type in its header. CoAP Content Format Registry provides the following initial entries: plain text, XML, JSON, EXI, octet stream, and link-format. New Internet media types may be used depending on the target IoT segment. COAP uses DTLS1.2, and security keys generated by DTLS are used to protect CoAP messages.

11.6   MQTT Application Protocol

The *Message Queuing Telemetry Transport* (MQTT) protocol supports IoT and M2M communication among locations/devices in networks. With small code footprints (which can run on 8-bit microcontrollers, 256 kb RAM-based devices), low power, low bandwidth, high-cost connections, high latency, variable availability, and negotiated delivery guarantees, MQTT is yet another protocol is used for large IoT-based networks. MQTT was initially known as SCADA protocol, which is suitable for sensor modules, medical devices, and home automation devices communicating to server/broker devices via satellite links. MQTT is suitable for mobile applications because of its small code size, smaller data packets, and efficient distribution of information to one or many receivers (subscribers). MQTT was invented by IBM and Arcom (now Eurotech) in 1999 and is approved by the Organization for the Advancement of Structured Information Standards (OASIS) as an open standard

Fig. 11.13 CoAP protocol stack

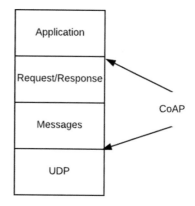

in the year 2015. In 2016, it was accepted as an International Organization for Standardization (ISO) standard. The protocol is continuously improving, and it now supports WebSocket, which is the real-time protocol for full duplex client-broker configuration. Later, notable versions included the v3.1.1 standard and v5.0 standard, both having been approved as OASIS standards. Latest version 5.0 supports better reporting of errors, metadata in message headers, shared subscriptions, message and session expiries, and topic aliasing.

Subscribe/Publish Scheme: The MQTT protocol shown in Fig. 11.14 is based on the principle of publishing messages and subscribing to topics (resources and services) or "pub/sub." Multiple clients connect to a server/broker devices which host many resources and services. Resources and service details are stored in topics or containers. The devices can subscribe to topics that they are interested in. Clients connect to the broker and publish messages to these topics or containers. Many clients may subscribe to the same topics. A publisher may publish a message once which can be received by multiple subscribers. Topics are structured into topic trees, which are treated as hierarchies, using a forward slash (/) as a separator. This allows arrangement of common themes to be created. Topics and topic trees can be created administratively, although it is more common for a server to create a topic on-demand (subject to security policies) when a client first attempts to publish or subscribe to it. A subscription may contain special characters, which allow clients to subscribe to multiple topics at once, within a single level or within multiple levels in a topic tree. The topics/containers are stored in broker devices to which the client and servers publish and subscribe.

Quality of Service (QOS): MQTT supports three levels of QOS depending on the capability of underlying TCP layer functionality. The three levels are QOS 0, QOS 1, and QOS 2. The higher level of QOS is more reliable but may have higher latency and data overhead. The definitions of these three types of QOS are as follows:

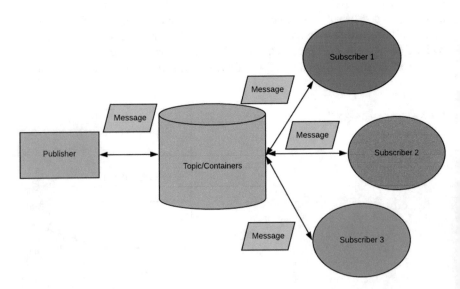

Fig. 11.14 Publish/subscribe scheme of MQTT clients and servers

QOS 0: MQTT client/server can communicate to broker device with messages sent without expecting the acknowledgement. This means that there is no guarantee that the message is received by the receiver. This is QOS 0 service level of quality.

QOS 1: This is acknowledged service level of quality. MQTT device will send the message to broker device and expects an acknowledgement on the reception of the message at the receiver. This results in at least one successful communication.

QOS 2: QOS 2 is assured service. MQTT device will send the message at least once to the broker device such that it reaches it before it gets deleted in the message queue. Figure 11.15 shows the message sequence chart in QOS 2 in cases of successful transaction and failed acknowledgement. The transactions in red color in Fig. 11.15 are failed transactions in first attempt but succeeded subsequently.

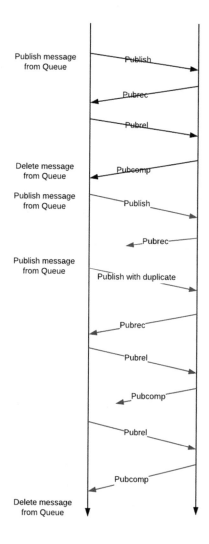

Fig. 11.15 QOS 2 level of communication service in MQTT

11.6.1   Retained Messages

Publish messages may be set to be retained. This means that the server/broker will keep the message even after sending it to all current subscribers. If a new subscription is made that matches the topic of the retained message, then the saved message will be sent to the requested client.

11.6.2   Durable and Non-durable Sessions

MQTT clients can request two types of sessions for message transfers. They are durable and non-durable sessions. Client requests non-durable session when both the devices discard previous state or message information when they get disconnected. In this case, every time, new subscriptions are requested by the devices. Clients can also request durable session wherein they need not resubscribe to the service when the new session is re-initiated. When the client discontinues any subscription, previous transactions will remain stored until it connects again.

11.6.3   Will Message

When a client connects to a broker, it may inform the broker that it has a Will. This is a message that it wishes the broker to send to interested parties when the client disconnects abnormally. The Will message has a topic, QoS, and retain status just the same as any other message. Abnormal disconnection can occur because of an I/O error in either of the server/broker during communication with the client, or the client fails to communicate within the Keep Alive timeout.

11.6.4   Protocol Stack

MQTT requires an underlying network protocol TCP and IPv6 that provides ordered, lossless, bi-directional connections. MQTT implementations use one or more of the following underlying protocols:

- Simple TCP/IP
- TCP/IP with Transport Layer Security (TLS)
- WebSocket with TLS or Websocket without TLS

The protocol stack is shown in Fig. 11.16.

MQTT protocol supports security with usernames and passwords in connection requests. Connections can be refused due to a bad username or password.

Fig. 11.16 MQTT protocol stack

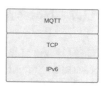

11.6.5 MQTT Tools

There are many tools and scripts available online for testing protocol. MQTT protocol testers are available for Windows, Linux, Android, and iOS. One can download these depending on the operating systems.

11.6.5.1 Command Line Tools

- Browser tools.
- Smartphone and tablet tools.
- These tools can be executed as command line options on a terminal or on a smartphone or a tablet.

Command Line Tools

These tools are part of the **Mosquitto broker** which can be installed on Windows or Linux or macOS.

Mosquitto_sub—Simple subscribe utility for testing. This comes with the Mosquitto broker.

Mosquitto_pub—Simple publish utility for testing. This comes with the Mosquitto broker.

Browser Tools: Chrome

MQTT-Lens—**Chrome add-on** for publishing and subscribing to an MQTT broker.

MQTT-Box—**Chrome add-on** for publishing and subscribing. It has more features than MQTT-Lens. It supports WebSocket.

Android Tools: Smartphone and Tablet

MQTT Dashboard—Simple tool for publishing and subscribing.

MQTT Dash—Create MQTT dashboards for the smart home.

SYS and **MQTT-monitor** are the topic monitor which can be used to monitor the topics on the MQTT broker.

11.7   Notes on Standard Defining Bodies

ETSI is one of the three European Standards Organizations (ESO) which define the standards for information and telecommunication for interoperability. The standards defined by ETSI are used worldwide.

ETSI-IoT radio layer standardization is carried out by **3GPP (3rd Generation Partnership Project)**. This defines functional inter-operability standards for **LTE-M, NB-IoT, and EC-GSM-IoT. 3GPP™** is a partnership project to bring all the national Standards Developing Organizations (SDOs) to define standards for third generation of mobile and cellular telecommunications and Universal Mobile Telecommunications System (UMTS). 3GPP now encompasses seven Standards Developing Organizations ETSI (Europe and the rest of the world), ATIS (USA), ARIB and TTC (Japan), TTA (South Korea), CCSA (China), and TSDSI (India).

ETSI-IoT service layer standardization is being carried out by **oneM2M**. oneM2M is the global standards initiative that covers requirements, architecture, API specifications, security solutions, and interoperability for machine-to-machine and IoT technologies. oneM2M specifications provide a framework to support applications and services such as the smart grid, connected car, home automation, public safety, and health. oneM2M is adopted as a national standard by **Telecommunications Standards Development Society, India (TSDSI)**, and is being used for developing 100 smart cities as a part of digital India program. The standard defines common service layer as shown in Fig. 11.17.

The service layer manages the following functionalities:

- Registration
- Discovery
- Security
- Group management
- Communication management
- Data management and repository
- Subscription and notification
- Device management
- Application and service management
- Network service exposure
- Location
- Service charging and accounting
- Transaction management

The **Open Connectivity Foundation (OCF)** [1] is establishing a single solution that addresses interoperability across multiple vertical markets like automotive, smart home, and healthcare by defining an efficient and scalable communication stack. OCF has based their specification on standard deployed technologies like REST and secure CoAP. OCF has come out with an optimal code which can run on

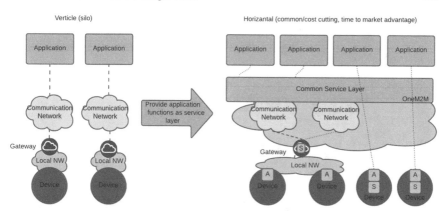

Fig. 11.17 OneM2M defines common service layer which will be used across applications. (Courtesy: IoT-Now/KETI)

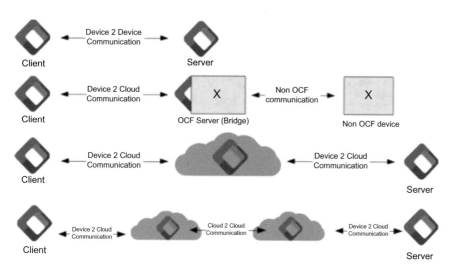

Fig. 11.18 OCF communication mechanism (Source: https://openconnectivity.org/technology/core-framework/)

device with small payloads in CBOR, JSON, and XML, using CoAP, for services depending on applications. The OCF communication mechanism is shown in Fig. 11.18. OCF definition covers device-to-device and device-to-cloud communication.

OCF framework solution space is shown in Fig. 11.19.

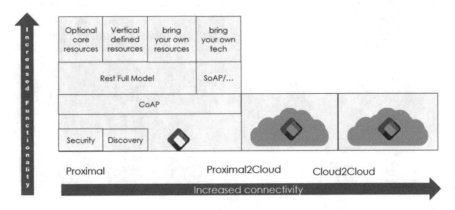

Fig. 11.19 OCF framework solution (Source: https://openconnectivity.org/technology/core-framework/)

References

1. https://www.etsi.org/deliver/etsi_gr/IP6/001_099/008/01.01.01_60/gr_IP6008v010101p.pdf
2. ETSI GR IPV6 008 V1.1.1(2017-06)
3. rfc791.pdf

Chapter 12
M2M Communication in Constrained Devices

12.1 Constrained Network and Constrained Devices

Devices which operate within close proximity or in a neighborhood and are connected to other devices are termed as **neighborhood area networks (NANs)** or **smart object networks** or **low-power and lossy networks (LLNs)**. LLNs are low-power, low-bandwidth networks, which can be wired or wireless. The data rate in this network is of the order of a few hundreds of kilobytes per second. Devices, which form these networks, can be **constrained devices**. **Constrained devices** are small smart devices working with 8-bit microcontrollers and are connected to a limited RAM/ROM. The networks formed by these devices are mostly connected in **star** or **mesh** network to ensure proper coordination. In these types of networks, the packet delivery ratio is typically 60%–90% with burst errors and occasional loss of connectivity. These networks and devices are simple, low power, easy to implement, reliable, low data rate, self-organizing, and low cost. Self-organizing means they form network on their own by connecting with neighboring devices like Bluetooth devices. Examples of LLN technologies are wireless IEEE 802.15.14g and Power Line Communications (PLC) such as IEEE 1901.2a networks. These can be found in a variety of application areas like home awareness systems with home keys, reminders, alarms of different connected appliances, light controls, water level controls, remote controllers. This network is also suitable for consumer appliance, energy sensing and control systems, asset monitoring systems for industries, safety sensors, and actuator-based devices.

© The Author(s), under exclusive license to Springer Nature Switzerland AG 2021
V. S. Chakravarthi, *Internet of Things and M2M Communication Technologies*,
https://doi.org/10.1007/978-3-030-79272-5_12

12.2   Internet Access

Constrained device on LLNs can access the Internet with direct connectivity or through gateways. The constrained devices form personal area network (PAN), and a device on PAN can connect to another device via one or more M2M gateway devices on the Internet. The M2M gateway acts as a **proxy** for the network domain devices. **Proxies** manage procedures of authentication, authorization, configuration, and provisioning of the devices. Examples of M2M area networks adopt technologies such as IEEE 802.15.1 Zigbee®, Bluetooth®, IETF ROLL, ISA100.11a, etc. or on local networks such as PLC, M-BUS, Wireless M-Bus, and KNX.

12.3   M2M Gateway Device

M2M gateway device runs on M2M application(s) software with M2M service capabilities and may sometimes are just sensor devices which are hidden from the network.

12.4   Network Domain

The network domain is composed of the following elements.

12.4.1   Access Network

The access network allows the M2M device and the gateway to communicate with the core network. Access networks include (but are not limited to) xDSL, HFC, satellite, GERAN, UTRAN, eUTRAN, W-LAN, and WiMAX.

12.4.2   Core Network

The core network supports internet (IP) connectivity at a minimum and potentially other connectivity means, such as:

- Service and network control functions.
- Interconnection with other networks.
- Roaming support across different networks which offer different feature sets.
- Core networks (CNs) include (but are not limited to) 3GPP, ETSI TISPAN, and 3GPP2.

12.4.2.1   M2M Service Capabilities

M2M service capabilities provide functions that are needed for machine to machine communication for different applications, which deal with the user or the environment. They use core network functionalities and expose these functions to other interfaces through a set of open interfaces. They hide network dependencies to optimize application development.

12.4.2.2   M2M Applications

These are application software for accessing M2M service capabilities via an open interface.

12.4.2.3   Network Management Functions

These consist of software functions for device provisioning, supervision, fault management, etc. which are required to access the core network.

12.4.2.4   M2M Management Functions

It consists of all the functions required to manage M2M service capabilities in the network domain. The management of the M2M devices and gateways use specific M2M service capability.

12.5   Personal Area Network (PAN)

Personal area networks (PANs) are small networks wired or wireless that operate within a short range and are suitable for in-premise networks and office network. The number of devices connected to PAN is less than hundred and sometimes tens. With the advancement of wireless technologies offering reliability close to wired networks, almost all PANs use wireless technologies. Wireless PANs are used for M2M communication in IoT landscape, because of their adaptability and ease of use. Use of a particular wireless PAN technology depends on the application, the range of operation, and the number of devices forming the network. Many new technologies like LoRa/LoRaWAN are still emerging, and low-bandwidth links, such as low-power wide-area network (LPWAN), need to be optimized to access the Internet. It is difficult to scale-up the IoT network forming PAN. The following discussion summarizes different wireless technologies for PAN:

IEEE 802.15.4 WPAN: The IEEE 802.15 standard is defined as a low data rate solution for a very long life, lasting many months to many years, operating on batteries in an unlicensed, international frequency band. Potential applications are sensors, interactive toys, smart badges, remote controls, and home automation.

Other well-known technologies used for PAN are:

- IEEE 802.11 WLAN (wireless local area network)
- LPWAN (low-power wide-area network)
- Cellular networks (NB-IoT, 5G)
- Power Line Communications (PLC), a wired technology emerging as IEEE 1901.2a and useful for home networking over power lines

12.6   Constrained Device and Its Architecture

Most of the sensor nodes are constrained devices. The capabilities of the sensor nodes are application dependent. A lot of factors like size, cost, and energy consumption play a role in deciding the device features. Typical architecture of the constrained device is shown in Fig. 12.1.

Any constrained device for a sensor node consists of the following modules:

Microcontroller: It is the main computing and processing module, which receives data and processes it for storage or communication. The processor can be simple an 8-bit or a 16-bit microprocessor. The data processing functions are implemented in a combination of hardware and software. The software for these devices is embedded in the on-chip memories, which could be operating system based or bare metal. Operating system is the execution environment, which is

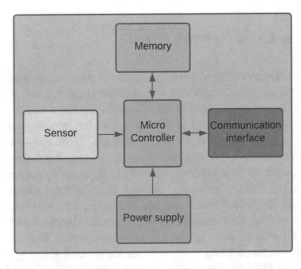

Fig. 12.1 Architecture of a constrained device

optimized for the firmware to operate in constrained environment. These micro-controllers support sleep modes, which wake up only after the sensor senses a parameter. For example, in a battery-operated WLAN-enabled device, many low power modes are implemented to obtain low power consumption wherein only when the received signal strength indicator (RSSI) is detected does the processor wakes up.

Sensor/actuator: This module interfaces with the external environment, which could be the user or the ambience to capture physical parameters. This may also contain the actuator, which can control the physical parameter.

Communication module: It enables data communication with other devices or machines over a air as communication channel in wireless technology. In wired communication, the interface can be CAN bus, Profibus, or LON bus, which are typical choices.

Memory module: Memory module will be of small size just to hold and forward the data onto the communication interface.

Power supply module: It can be a battery or power supply with the supporting circuitry. Each of the modules has to trade off between power and data/signal processing. Many times, these devices support multiple power modes and unused modules. Using appropriate power modes, unused modules are swtched off while executing a particular function to achieve low power consumption.

12.7 Wireless Sensor Network (WSN)

Wireless sensor networks are of two classes: mobile ad hoc networks (MANET) and fieldbuses. These networks meet most of the requirements of small networks formed by constrained devices and hence are the preferred choice for sensor networks. Characteristic features of the WSN are **type of service and quality of service** with acceptable latencies, promised data rates, fault tolerance in spite of association and disassociation of sensor nodes, scalability, dynamically changing number of devices on the network, programmability, and easy maintainability. Mobile ad hoc network (MANET) is a network formed to communicate the data immediately. Hence, the network is configured as and when required. Devices on WSN interact with the environment for sensing physical parameters and support low data rate and burst data traffic during communication. **MANETs** are more conventional network to support applications like web access and voice/multimedia communication.

Fieldbuses are networks designed for operation under real-time constraints with built-in fault tolerance and are well suited for control applications. Examples are Profibus and IEEE 802.4 token bus networks for factory automation or CAN bus for automobiles.

The WSN technologies are based on the following standards.

12.7.1   Smart Utility Network (SUN)

SUNs are formed by low-power simple devices which can execute multiple applications based on IEEE 802.15.14g standard. They can co-exist with other technologies operating in the same range of frequencies. Some SUN devices are designed to operate in very-large-scale, low-power wireless applications requiring large transmit power within the frequency spectrum permitted by the regulations, to achieve long-range and point-to-point connections. Frequently, SUNs are required to cover geographically widespread areas containing a large number of outdoor devices. In these cases, SUN devices are able to employ mesh or peer-to-peer multi-hop techniques to communicate with an access point.

12.7.2   Rail Communications and Control (RCC)

Devices based on RCC standard are used for exchanging information between different automobiles, trains, and other mobile rolling stock to fixed trackside or network infrastructure wirelessly. RCC devices are intended to support mobile rail vehicle communications at high speeds for a range of over 50 km.

12.7.3   Television White Space (TVWS)

TVWS operations apart from data communication need to have a way to access the available channel for communication which is saved as a centralized database on the Internet. This means that these devices should have Internet access to this database if they do not have access to another TVWS device with Internet access through which it can determine the available channel information.

12.7.4   Radio Frequency Identification (RFID)

This technology is used for the identification of an object or a person with simple exchange of information using radio frequency communication at short-distance or near-field communication (NFC). There are two types of RFIDs: active and passive. Active RFID devices (typically called tags) need to be powered for its operation, while passive devices get activated or enabled when it comes closer to an RFID reader which transmits signals periodically. Typical applications include asset management, inventory management, people/material tracking, process control and automation, safety, and accountability. In its simplest form, an active RFID system comprises a number of transmit-only tags that periodically transmit a packet containing a unique ID and a small amount of data. This will be received by the RFID readers located in fixed places. When the device with the tag comes near the receiver, the communication identifies the device and processes based on applications. More complex active RFID systems might employ two-way communications with the tag for control, communication, and coordination.

12.7.5   Low-Energy, Critical Infrastructure Monitoring (LECIM)

This part of standard defines LECIM system infrastructure, which is suitable for a large number of devices which are not powered from the main supply and are distributed across large areas and are widely spread across. The devices can be located at the challenging environmental conditions, and they can be programmed to send the data at pre-scheduled times. Low power modes are defined such that they wake up just for the scheduled times and the remaining times they are in sleep modes and hence do not consume power. These networks can operate for years. Applications where LECIM network solutions are suitable are periodic environment monitoring, traffic monitoring, infrastructure monitoring, oil and gas pipeline monitoring, electric net metering, rail/road condition monitoring, bridge/structural integrity monitoring, water leak detection/sewer monitoring, person tracking, cargo container tracking, perimeter security/border surveillance, etc.

12.7.6   Medical Body Area Network (MBAN) Services

MBAN devices are designed to be deployed within a hospital setting. MBAN devices provide a cost-effective way to monitor patient in real time, so that clinicians can intervene to provide care when needed. Wireless devices that operate on an MBAN spectrum are disposable sensor devices which are worn by the patient, which monitor a patient for vitals like blood glucose and blood pressure, delivery of electrocardiogram readings, and even neonatal monitoring systems. MBAN technology will make it easier to move patients to different parts of the healthcare facility for treatment and can dramatically improve the quality of patient care by giving healthcare providers a chance to identify life-threatening problems or events before they reach critical levels. These devices share the spectral band (reserved for only medical use) in an orderly manner which is pre-configured. These devices operate in the reserved bandwidth of 40 Mhz of reserved spectrum in the 2360–2400 MHz band and operate in a short range of hospital room like Bluetooth technology with limited data rate of less than 2Mbps. Typical MBAN system is shown in Fig. 12.2.

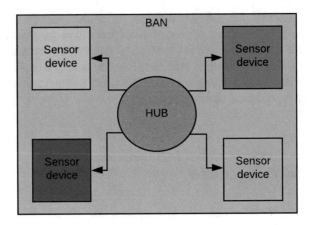

Fig. 12.2 MBAN system

12.8   Multi-PHY Management (MPM) of the SUN WPAN

Multiple and different SUN PHYs can operate at the same location and within the same frequency band. In order to mitigate interference, an MPM scheme is specified for smart utility network (SUN) to facilitate inter-PHY co-existence, interoperability and negotiation among potential coordinators with different PHYs. Potential coordinator detects an operating network during its discovery phase using the common signaling mode (CSM) appropriate to the frequency band being used. The MPM procedure can be used in conjunction with the clear channel assessment (CCA) mechanism to provide co-existence.

12.9   Network Architecture of WPAN

WPAN systems consist of many components. The simplest device consists of at least one IEEE 802.14-compliant MAC and physical layer components with transceiver and radio interface. WPAN is made of two or more of 802.14 devices. The network will have one full featured device and one or more radio frequency device (RFD). The layered architecture of a simple WPAN system is shown in Fig. 12.3. Following are few terms used in understanding the WPAN technology:

PD-SAP: Physical device service access protocol
PLME-SAP: Physical layer management entity service access protocol
MLME-SAP: MAC layer management entity service access protocol
MPDU-SAP: MAC protocol data unit-service access protocol

Fig. 12.3 WPAN layer
architecture

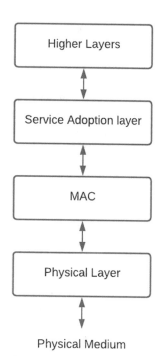

The functions of different layers are as follows: The main function of the *physical component* are to enable-disable the transceiver, receive signal detection, assessing channel quality, and transmitting and receiving the signals. The function also enables co-existence of other devices in the network without much of interference.

The *MAC* layer functions include:

- MAC data service enabling transmission and reception of data packets, called MAC layer data unit (MLDU) services.
- MAC layer management entity service access protocol function which manages link establishment by association and disassociation procedures.
- Keeping the link active.
- Processing the data by validating it for data correctness by CRC or FCS.
- Acknowledging the reception of the MAC protocol data unit (MPDU). It also provides security function by supporting encryption and decryption of the MPDUs.

12.10   Routing Protocols for Sensor Networks

In traditional internet systems based on Ethernet protocols, routing functionality is implemented in layer 2 which is the data link layer. This architecture is convenient when the network is formed by most reliable, stable networks. In PAN or short-range networks formed by IoTs or IIoTs, it is difficult to implement routing functionality on constrained hardware resources. These devices are of different types complying to different protocols, meaning, the networks are formed by the sensor devices like PLCs, FANs, and sometimes wearable devices. These devices need routing function to be flexible, reconfigurable, and dynamically adaptable. This is achieved by implementing routing at the network layer of the device systems. Considering this, IETF has defined the "Routing over Low-Power and Lossy Networks" (RoLL) protocol standard which is an IPv6 routing protocol for constrained large-scale networks. This is used for implementing devices for applications in urban networks, like smart grid, industrial, and home building automation networks. For vast number of the sensor network applications, IETF has defined a new standard called IPv6 routing protocol for *low-power* and lossy (RPL) networks. RPL provides support for a large number of technologies and features that matches all service requirements. RPL is highly flexible and supports multiple protocols which can be designed to operate in harsh conditions. Many technologies like Zigbee follow routing at the network layer as defined in RPL.

12.11   Constrained Application Protocol (CoRE)

CoRE is a framework for resource-oriented applications intended to run on constrained IP networks. A constrained IP network supports small packet sizes, exhibits a high degree of packet loss, and has a substantial number of devices which can be turned ON/OFF on need basis. These networks have low throughput, consume low power, and work on constrained hardware. The network device uses limited software code size and a small RAM per node. Home and building automation networks and energy management systems are examples of core networks. These networks comply with Constrained Application Protocol (CoAP). CoRE architecture is based on RESTful architecture as shown in Fig. 12.4. The network shown in yellow background is the CoRE network where all the devices are constrained devices. The conventional IoT network is shown in blue in Fig. 12.4.

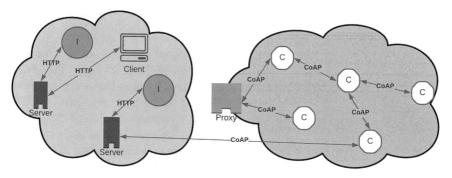

Fig. 12.4 CoRE architecture

12.11.1   6LoWPAN

6LowPAN is based on IEEE 802.15.4 standard which is LoWPAN on IPv6. 6LoWPAN supports host-initiated neighbor discovery to find the devices existing in the vicinity. The data communicated on this network among devices are very low data rate packets. Since they operate in noisy environments, techniques like header compression are used to protect the data packets. There is an **service adoption layer** added on IPv6 where this is done. Since the protocol is built on IPv6, it can support a large number of sensor devices.

12.11.2   Working Group 6Lo

6Lo has defined the adaptation of this protocol on data link layer (layer 2) technologies like Bluetooth. 6Lo working group also defines support for security and other functions for 6LoWPAN on constrained networks. Running IPv6-on-IoT guarantees application portability across devices, scalability, and manageability in the Network using standard IP protocols.

12.11.3   Bluetooth Low Energy (BT-LE)

BT-LE specifies the IPv6 over Bluetooth Low Energy (BT-LE). The Bluetooth standard has been widely implemented and available in mobile phones, notebook computers, audio headsets, and many other devices. The low-power version of Bluetooth is a specification specially defined for sensors, smart meters, appliances, etc. BT-LE is a low data rate protocol which can even cover a long range in kilometers.

12.11.4 DECT Ultra Low Energy

The transmission of IPv6 packets over **Digital Enhancement Cordless Telecommunications** (DECT) Ultra Low Energy (DECT-ULE) enjoys the advantages of parent DECT technology and is customized for ultra low energy systems. The original DECT technology is used worldwide for cordless telephony. Over 600 million homes deploy cordless telephony for home cordless telephony. DECT modified for ultra-low energy enjoys features such as long-range, worldwide reserved frequency band and interference-free communication. The technology is also used for sensors, smart meters, and home networking devices.

12.11.5 Z-Wave IETF

Z-Wave is a wireless communication protocol used for home automation. Home appliances are connected using this protocol in mesh configuration. The Z-Wave standard describes the frame format for the transmission of IPv6 packets as well as a method of forming IPv6 connections, local addresses and method to generate stateless self-configuring IPv6 addresses. Z-Wave is primarily used for at-home devices.

12.11.6 Power Line Communication (PLC)

IPv6-based PLC is used for industrial process automation. It uses IPv6 packets over IEEE 1901.2a Power Line Communications (PLC) technology, specified in ITU-T standards. IEEE1901.2a supports layer 2 routing which is required for machine automation because of the real-time need of IIoTs.

12.11.7 G3-PLC

G3-PLC is an open standard network for smart meters and other low-power electrical devices, which is also used for electric grid automation. It is hoped that in the future Power Line Communication is supported using this standard.

12.11.8 Near-Field Communications (NFC)

NFC technology is used for communication between two smart devices for instant file transfers. This specifies the transmission of IPv6 packets over the NFC L2 technology which is a very-low-range (~10 cm) communication. It uses techniques like IPv6 header compression, address formation, and neighbor discovery optimizations for this short range but is also useful for many social and home applications via smartphones and other devices.

12.11.9 BACNET Master-Slave/Token-Passing (MS/TP)

Buildlng automation control network (BACNET) is another standard for building automation and control. It also supports multicast communication. The standard defines a medium access control method for the RS-485 physical layer, which is used extensively in building automation networks. IETF standard **6lobac** defines the IPv6-based addressing mechanisms and transmission of packets over MS/TP networks.

12.11.10 802.15.4e Time Slotted Channel Hopping (TSCH): The IEEE 802.15.4e

Time Slotted Channel Hopping (TSCH) is an amendment to the medium access control (MAC) portion of the IEEE 802.15.4 standard for industrial automation and process control LLNs. IPv6 over TSCH also uses 6lo/6LoWPAN protocols. This is used for process automation systems.

12.11.10.1 IEEE 802.11ah

The IEEE 802.11ah standard defines a WLAN system operating at sub 1 GHz license-exempt bands designed to operate with low-rate/low-power consumption. This amendment supports a large number of stations and extends the radio coverage to several hundreds of meters. IEEE 802.11ah technology presents a trade-off between energy consumption and bit rates. Thus, it is beneficial to run a 6lo-defined IPv6 specification in order to save energy in the packet transmission in the IEEE 802.11ah-supported topology, stateless address auto-configuration, and neighbor discovery. Applications of IEEE 802.11ah range from smart meters, appliances, home devices to the industrial applications/monitoring devices.

12.11.10.2   Low-Power Wide-Area Network (LPWAN)

Low-power wide-area network (LPWAN) is a type of wireless telecommunication network designed to allow long-range communications at a very low bit rate among battery operated devices.

The some of the major technologies in this range are the following:

- LoRaWAN
- Cellular LTE-based technologies (defined by 3GPP):

    - LTE-MTC
    - NB-IoT
    - EC-GSM
    - IoT

- UNB (ultra narrowband)
- WI-SUN
- Sigfox

12.11.10.2.1   LoRaWAN

LoRaWAN is a protocol for a high-capacity, long-range, and low-power network that the LoRa Alliance has standardized for low-power wide-area networks (LPWAN). The LoRaWAN protocol is optimized for low-cost, battery-operated sensors and includes different classes of nodes to optimize the trade-off between network latency and battery lifetime.

12.11.10.2.2   Cellular LTE

This network is bi-directional and power-efficient and provides extended coverage via existing cellular networks. LTE Cat M1 and NB-IoT are cellular technologies designed for low data rate applications and requiring low power consumption. Some typical applications include metering, environmental and industrial monitoring, object location tracking, e-health, smart cities, smart agriculture, wearables, and many others.

12.11.10.2.3   Sigfox

Sigfox is a cost-effective, IoT-dedicated network that guarantees a high level of service and security for years to decades of low-power battery life for connected devices, in all business and industry sectors.

12.11.11   Industrial Automation and the Internet

Industry automation poses different kinds of challenges unlike the traditional Internet. They have distinct communication requirements that have to be considered for automation [1]. Industry automation can be classified as:

1. Building automation
2. Process automation
3. Factory automation
4. Substation automation

Well-known industry automation pyramid [1] is shown in Fig. 12.5. It is the layer architecture of industry automation along with the timing criticality required for automation at different layers. Layer 1 is called field network which has sensors and actuators which directly interact with the machines and processes. This layer is very critical and is characterized by low latency, real time, and zero tolerance toward failures as any mistake in this layer may cause huge damage endangering persons in the shop floor. Layer 2 is the control network consisting of servers and controllers whose functions are critical for machine floor operations. The top layers in the architecture are the operational network used for monitoring the processes and machines, which consist of monitoring servers and human-machine interfaces with operator's dashboards. The highest layer is the business management layer, which is the decision-making layer. The bottom two layers are **operations technologies (OT)** directly impacting process and machines in operation. Layer 3 onward consist of the traditional **information technologies (IT)**. The top layer will only be

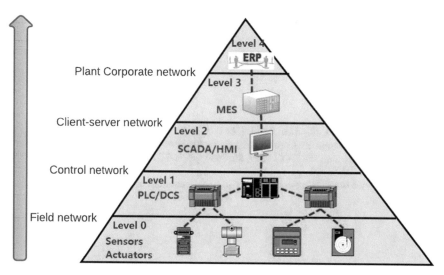

Fig. 12.5   Automation pyramid [1]

connected to the Internet for remote access and management. The major difference between conventional networks and industrial networks lies in the requirements of operational technologies, which are near real time (not in the conventional sense) and demand reliable control with criticality of timing as they directly control machines. Concepts like retransmission do not work here as the control messages for controlling the machines and processes have to be correct and have to be at the machines at the right time. The requirement is considered time critical and is a hard requirement. The automation designs are targeted to work on f (M, K) tolerance meaning the machines are designed to work M out of K control attempts which eases the requirement of the control system to an extent. Predictability and determinism is another property of OT layers, which is required to have zero downtime of the machines and processes. Safety is absolute requirement of the machine automation systems as the damage caused can be fatal to the business and persons if automation goes wrong. Another requirement in OT layers is the security threat. Any breach in security is like losing control over the manufacturing.

PLCs with proportional integral derivative (PID) are typically used for control functionalities of the first two layers in industry automation systems. The criticality of timing, low latency, and reliability are the major considerations in such systems. The IIoT systems for factory automation therefore involve complex, multi-layered architectures consisting of field I/O networks; remote terminal units (RTUs), programmable logic controllers (PLCs), and programmable automation controllers (PACs); and communication servers and supervisory control and data acquisition (SCADA) hosts in these lower layers of the architecture. Automation systems need wireless deterministic networking with fully scheduled radios such as TSCH modes of IEEE 802.15.4 and LTE/5G with the correct set of security technologies. Both ISA100.11a [2] and wireless HARTTM use variations of the TSCH MAC, which is optimized for ultra-low-power activities and is a natural match to transport low-frequency periodic flows, such as control loops, over a fully scheduled network.

References

1. Lennvall, T., Gidlund, M., Åkerberg, J. (2017) "Challenges when bringing IoT into Industrial Automation In: Darryn R. Cornish (ed.), 2017 IEEE AFRICON: Science, Technology and Innovation for Africa, AFRICON 2017, 8095602 (pp. 905-910). IEEE https://doi.org/10.1109/AFRCON.2017.8095602
2. https://www.isa.org/standards-and-publications/isa-standards/isa-standards-committees/isa112

Chapter 13
IoT Database Management and Analytics

13.1   IoT Database Analysis Framework

With thousands of IoT-based smart devices connected, there will be a large amount of data generated and communicated continuously on the Internet which has to be organized, stored, and analyzed to derive meaningful information or to control the physical parameters surrounding it. A new database management framework is needed because of the flexibility and spatial diversity requirement in the systems on the Internet for IoT applications. The **IoT data management framework** has to be a dynamic and a distributed network infrastructure, in which the device both physical and virtual entities are identifiable, autonomous, and self-configurable. Things should have M2M connectivity in addition to sense and control function of the environment in which they operate. The IoT platform should have scalable resources, services, and applications that are distributed and shared. This includes a large amount of data processed at different subsystems in the network and centrally stored which could be accessed whenever any subsystem needs them in full or in part. This data-based IoT framework is bound to enable many innovative **value-added services** with analytics, data sets, or applications to the stakeholders and provide various opportunities of collaboration in service offerings.

   Data management framework refers to architectures, communication protocols, and procedures for proper management of the data distribution. In the context of IoT, data management can also be considered as a layer between the devices generating the data and the applications accessing the data for analysis and services. The devices are subsystems in the data management framework. The functionality and data provided by these subsystems are to be made available to the IoT network, depending on the level of privacy desired by the subsystem owners accessing them. **IoT data** is a massive heterogeneous, streaming, and geographically dispersed real-time data generated by millions of diverse devices. These devices send their observations about certain monitored parameters, and processes report the occurrence of normal or abnormal events. Data from continuous monitoring are the most demanding in terms of communication overhead and storage, while events like retrieve requests and querries pose time constraint with end-to-end response times depending on the urgency of the response required by the event. Furthermore, there is metadata that stores information about the things, which generate data, with device/object identification, location, processes, and services provided by it. IoT data will statically reside in fixed or dynamic databases and communicate across the network from mobile objects to centralized storage points. Communication, storage, and process will thus be the defining factors in the design of data management solutions for IoT. IoT data management involves communicating a variety of raw or partially processed data to form a database at a centrally located place as a different subsystem of the network, which could be a cloud server or a device or edge servers. This database needs to support Atomicity, Consistency, Isolation, and Durability (ACID) properties. In this framework, the **thing or device layer** is composed of all entities and subsystems that can generate data. Raw data, or simple aggregates, are then transported via a communication layer to data repositories. These data repositories are owned by either organizations or the public, and they can be located at specialized servers or on the cloud. Organizations or individual users have access to these data repositories via **query** and **federation layers** that process queries and Analyses tasks, decide which repositories hold the needed data, and use protocol to get the necessary data. In addition, real-time or context-aware queries are handled through the federation layer via a **sources layer** that seamlessly handles the discovery and engagement of data sources. The whole framework therefore allows a two-way publishing and querying of data. This allows the system to respond to urgent data and processing requests of the end users and provides archival capabilities for later long-term analysis and exploration of value-added trends.

13.2   Big Data Life Cycle

The life cycle of data within an IoT system is as shown in Fig. 13.1. Data life cycle starts from data production or capture to data aggregation, delivery, optional filtering, and pre-processing and finally to storage and archiving. Querying and analysis are the end points that initiate (request) and consume data production, but data production can also be at IoT device which consumes services. Production, collection, aggregation, filtering, and some basic querying and preliminary processing functionalities are considered online, communication-intensive operations. Intensive pre-processing, long-term storage and archival, and in-depth processing/analysis are considered as offline storage-intensive operations.

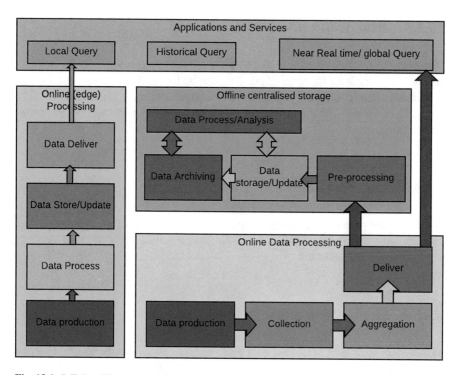

Fig. 13.1 IoT data life cycle and data management

13.2.1   Data Generation

Data is generated in devices or things of different types:

Passive device: The data generated by the devices which do not have a power source connected to it. Examples of such devices are RFID tags and ATM cards. These devices generate contactless identities or barcodes on them.

Active devices transmit their identity and status or environmental parameters captured by them. These devices are connected to active power supplies. Examples are smart streetlights and wireless sensor nodes.

Event data: This data is generated when devices receive certain triggers or events. The examples are motion sensors and fall sensors.

Real-time data is generated by devices periodically by configuration. These are periodic environmental parameter sensors, continuous body vital sensors, etc.

13.2.2   Data Acquisition

Receiving and storing data generated by the devices or received through M2M com-munication is called data acquisition. This can be initiated by configuration of the microcontrollers connected to the devices or triggered by a remote device. The application software or the device management software also can configure data acquisition.

13.2.3   Data Validation

Not necessarily the data captured is accurate and reliable. Therefore, data acquired is validated by applying certain rules, logic, and semantic annotations before they are stored or communicated.

13.2.4   Storage Operations

Database storage aims at making data available on a long term for constant access/updates, while archival is concerned with read-only data. Some of the IoT systems may generate, process, and store data within network for real-time and localized services. With no need to transfer this data further up to a centralized storage in the system or edges, both processing and storage elements may exist as autonomous units in the cycle. This is also called spatial storage of data depending on where it is relevant in the network.

13.2.5   Querying

Querying is the means to access and retrieve data from a database. In the context of IoT, a query can be either to request real-time data for temporal monitoring or to retrieve a certain view of the data collected. The first case is typical when a (mostly localized) real-time request for data is needed. The second case represents a more globalized view of data and in-depth analysis of data trends and patterns in part or full.

13.2.6   Production

Data production involves sensing of data by the sensor devices or things within the IoT framework and transmitting this data to interested parties periodically (as in a subscribe/publish model). It also involves sending data to the network aggregation systems locally or in the network and subsequently to database servers or sending it as a response triggered by queries that request the data from sensors and smart objects. Data sent is usually time-stamped and possibly geo-stamped.

13.2.7 Collection

Sensors and smart objects within an IoT sometimes store the data for a certain time interval or report it to host components. Data is collected at concentration points or gateways within the network where it is further filtered and processed and aggregated and compressed for efficient transmission. This is done in WPAN networks using wireless communication technologies such as Zigbee, Wi-Fi, and cellular.

13.2.8 Aggregation/Compression

Transmitting the raw data out of the network in real time is expensive given the increasing data streaming rates and the limited bandwidth. Aggregation and compression techniques are adopted to process the data in real time to make the data communication efficient.

13.2.9 Delivery

The data is filtered, aggregated, and, possibly, processed either at the concentration points or at independent virtual subsystems within the IoT. The results of these processes may need to be transferred to the system, either as final responses or for storage and in-depth analysis. Wired or wireless broadband communications is used to transfer data to permanent data stores.

13.2.10 Pre-processing

IoT data will come from different sources with varying formats and structures. Data may need to be pre-processed to add missing data, remove redundancies, and integrate data from different sources into a unified schema before being committed to storage. This pre-processing is a known procedure in **data mining** called **data cleaning**. At this phase, data is validated by adding probabilities at different levels in order to remove uncertainty that is present in data or to deal with the lack of trust in data sources.

13.2.11   Stored Data Update-Archiving

This phase is efficient storage and organization of data as well as continuous update of data with new information as it becomes available. Archiving refers to the offline long-term storage of data that is not immediately needed for the system's ongoing operations. The core ability of the centralized storage is the deployment of storage structures that adapt to the various data types and the frequency of data capture. **Relational database management systems** (RDMS) are a popular choice that involves the organization of data into a table with predefined interrelationships and metadata for efficient retrieval at later stages. Storage can also be decentralized for autonomous IoT systems, where data is kept at the objects that generate it and is not sent up the system. However, due to the limited capabilities of such objects, storage capacity remains limited in comparison to the centralized storage model.

13.2.12   Processing/Analysis

This phase involves the ongoing retrieval and analysis of stored and archived data in order to gain insights into historical data and predict future trends or to detect abnormalities revealed by the data that may trigger further investigation or action. Task-specific pre-processing may be needed to filter and clean data before meaningful operations take place. When an IoT subsystem is autonomous and does not require permanent storage of its data, but rather keeps the processing and storage in the network, then in-network processing may be performed in response to real-time or localized queries.

In Fig. 13.1, the flow of data takes one of the three paths:

1. A path for autonomous systems within the IoT that proceeds from query to production to in-network processing and then delivery
2. A path that starts from production and proceeds to collection and filtering/aggregation/fusion and ends with data delivery to initiating (possibly global or near real-time) queries
3. Finally, a path that extends the production to aggregation further and includes pre-processing, permanent data storage and archival, and in-depth processing and analysis

13.3   IoT Data Framework

IoT network needs a data framework for supporting data communication from a heterogeneous set of devices and networks, paving way for adaptation and seamless integration of other IoT subsystems. It requires reliable and comprehensive data management solutions that support interoperability between diverse subsystems and integrate the overall life cycle of data management and is context aware. The framework layers map closely to the phases of the IoT data life cycle shown in Fig. 13.1. The *"thing" layer* encompasses IoT sensors and smart objects (data production objects), as well as modules for in-network processing and data collection/real-time aggregation (processing, aggregation). **The *communication layer*** provides support for transmission of requests, queries, data, and results (collection and delivery). The *data/sources twin layers*, respectively, handle the discovery and cataloguing of data sources and the storage and indexing of collected data (data storage/archival). **The data layer** handles data and query processing for local, autonomous data repository sites (filtering, pre-processing, processing). **The *federation layer*** provides the abstraction and integration of data repositories that is necessary for global query/analysis requests, using metadata stored in the data sources layer to support real-time integration of sources as well as location-centric requests (pre-processing, integration, fusion). **The *query layer*** handles the details of query processing and optimization in cooperation with the *federation layer*.

The query layer includes the *aggregation sub-layer*, which handles the aggregation and fusion queries that involve an array of data sources/sites (aggregation/fusion). **The *application/analysis layer*** is the requester of data/analysis needs and the consumer of data and analysis results. The architectural layers proposed IoT data management framework and their respective functional modules are illustrated in Fig. 13.2. Figure 13.3 illustrates the functionality of each of these layers in data management framework with the relevant protocols.

13.4   Data Framework for WPAN

Most of the data management are targeted to WPANs, which are only a subset of the global IoT space, and therefore do not explicitly address the more sophisticated architectural characteristics of IoT. WPANs are a mature networking paradigm whose data management solutions revolve mainly around in-network data processing and optimization. Sensors are mostly stationery and resource-constrained nature, which does not facilitate sophisticated analysis and services. The main focus in WPAN-based data management solutions is to harvest real-time data promptly for quick decision-making, with limited permanent storage capacity for long-term usage.

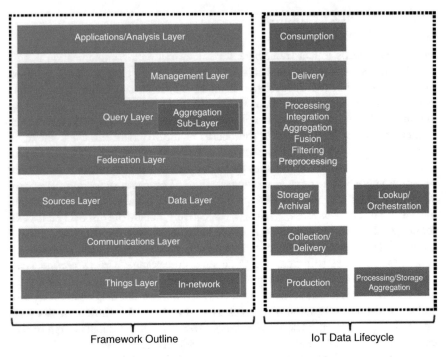

Fig. 13.2 IoT data management framework mapping to data life cycle (Courtesy: Sensors 2013, 13, 15582-15612; doi:10.3390/s131115582)

13.5   Industrial Data Life Cycle

Industrial data management is very crucial for Industry 4.0. Very important resource of industrial automation worldwide can be of huge value if managed properly. Managing this data requires high processing and storage capabilities due to its huge, complex, unstructured, and heterogeneous nature. The life cycle of industrial data can be defined with the help of three phases like physical, middleware, and application or services as illustrated in Fig. 13.4. The whole industrial environment can be divided into two subsections like the real and the digital world. Real-world environment is a smart factory where the IIoT-enabled machines produce process parameters which form the raw industrial data with various data types, formats, and diverse dimensions. The data sources in the physical devices are status trackers, process sensors, web-generated data, databases, and the data from third-party applications. These are digitized, accumulated, and aggregated from different factory environments and are transferred to the systems in digital world, where middleware and application layers offer different services to manage them. The middleware component addresses interoperability across various factory devices, device discovery, scalability, management of massive data, context awareness, and the security features of the IIoT environment.

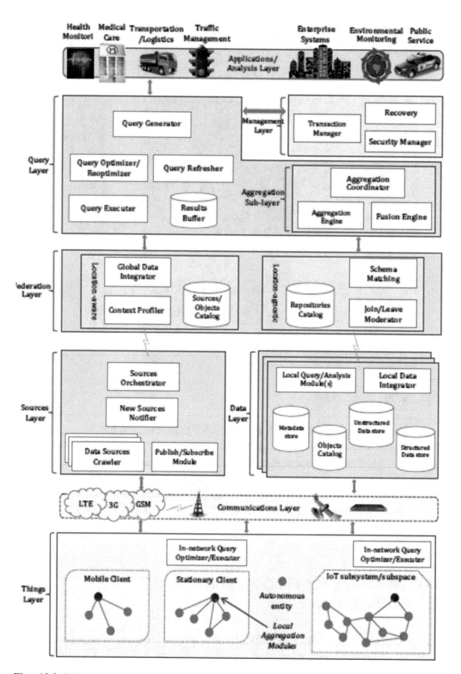

Fig. 13.3 IoT data management framework. (Courtesy: Sensors 2013, 13, 15582-15612; doi:10.3390/s131115582)

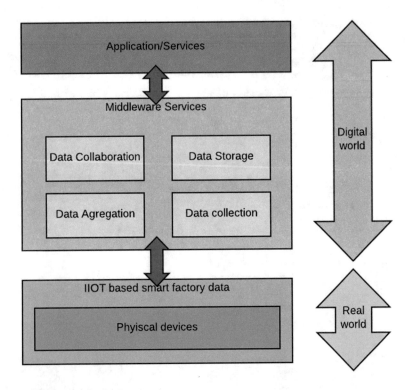

Fig. 13.4 Industrial data life cycle

In the middleware phase, the data collection module directly interacts with the physical devices and exchange (configuration/status and control) information using commands/response method. The data collected is processed by filtering, sorting, aggregation, and classification techniques and is sent to a permanent storage as data repositories for later use. This database, nowadays, is stored in cloud computing systems which are more reliable, secure, scalable, and robust.

13.6 Industrial Data Management System (IDMS) Framework

Figure 13.5 shows the five-layer IDMS framework for smart factories. In this framework, **the physical layer** contains all industrial sensors, actuator, and field devices, which are responsible for the creation of raw data and unique events. **The communication layer** implements the latest industrial protocols and makes sure of the secure connection between each layer of the system. This layer also controls

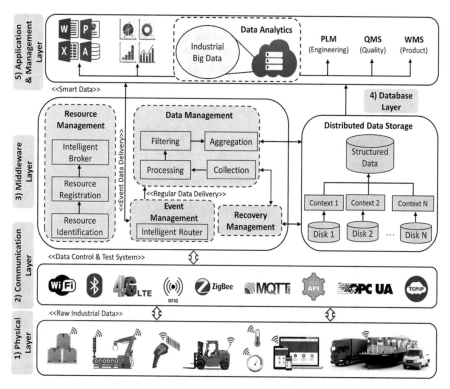

Fig. 13.5 Industrial data management system framework [3]. (Source: Framework of an IoT-based Industrial Data Management for Smart Manufacturing", Journal of Sensor and actuator, MDPI, 28 April 2019)

streaming of data, queries, requests, and results. **The middleware layer** consists of different functional components to provide support for discovery of diverse data sources and apply data processing processes. **The database layer** supported by the local repositories provides partial storage for distributed industrial data to overcome communication delays and intensive workload on the cloud server. **The application layer** handles the queries from consumers and provides real-time data analysis.

13.6.1 Physical Layer

The physical layer contains all data producer entities of the manufacturing system and modulus. These entities can be flow meters, servo meters, robots, conveyor belts, PLCs, machine visions, smart containers, embedded chips, and various devices on the shop floor. Real-time data acquired from the factory floor is transmitted to the upper layer by different adapters using communication protocol. These adapters correspond to different sensors such as temperature, pressure, energy, vibration, rotational speed, force, torque, and acoustic emission to monitor the real status of machines. All physical devices on the shop floor are integrated with each other, and every device has unique identification for easy access by the user. The local aggregation screens the acquired data by discarding the less important or redundant data streams.

13.6.2   Communication Layer

The communication layer provides human-machine interface, connects all layers of the proposed framework, and offers transmission links between data producers and data consumers. This layer also handles communication between distributed factory devices within a vast industrial area to concentrate data collection, processing, visualization, and storage.

The industrial data should be efficiently collected from the physical layer and transmitted with high throughput and low latency to upper layers for further processes and analysis. Various communication technologies such as RFID, Wi-Fi, Bluetooth, Wi-Fi Direct, 4G LTE, Z-Wave, Zigbee, etc. are being used to transfer heavy data traffic with minimum latency. Security is implemented in this layer. The common protocols (i.e., IPv6, MQTT, SOAP, REST API, Open platform communication unified architecture (OPC-UA) (for object linking and embedding-unified architecture)) are used for communication and control in this layer. The MQTT is used to acquire data from various devices and transmit to the middleware layer that needs to be controlled and monitored by the cloud server. The REST API protocol is specifically used for secure collection of data from IIoT.

OPC-UA has been defined as a standardized industrial communication protocol for the reliable, secure, and vendor-neutral transmission of raw data from the factory/process machine sensors.

13.6.3   Middleware Layer

Middleware layer offers various services for applications development by integrating heterogeneous computing and communication devices. This layer supports interoperability within diverse applications and services running on these devices. This layer directly interacts with data producer at the physical layer and data consumer at either the application layer or distributed data storage layer. The major functions of this layer are as follows:

13.6.3.1 Resource Management

The resource management component makes sure of the discovery and arrangement of data sources by storing the database fragment locations for updating and querying purposes. It typically manages **resource identification, resource registration**, and **intelligent broker** functions. *Resource identification* handles the continuous and real-time queries for the physical machine sources. Identity of new sources is verified by authorizing and passing the device specification to the metadata store. *Resource registration* is responsible for data source registry and provides functions between physical and application layers to make agreements for sharing of data and queries. *Intelligent broker* behaves as a mediator between data consumers and data producers. It supervises and keeps records of each data exchange transaction and provides the rollback in case of incomplete and faulty data exchange transactions. Data analysis and quality-related services are also provided by the broker.

13.6.3.2 Event Management

After collecting the raw industrial data through smart devices and the intelligent network, it is delivered either directly to the application layer for immediate responses to events requiring urgent action or to the data management component for pre-processing and storage of normal data stream.

13.6.3.3 Data Management

Data generated by factory devices is delivered to the data management component of the data distributed system (DDS) for pre-processing. The DDS can support multiple connections within IIoT environment simultaneously. It comprises two main functions: data collection and aggregation. During the data collection function, data is acquired from devices at the physical layer and is pre-processed and is sent to the aggregation component for summarizing into different chunks. Finally, the partially structured and aggregated data is sent to the distributed data storage module where it is fully structured by applying context extraction techniques. Typical data management flow is shown in Fig. 13.6.

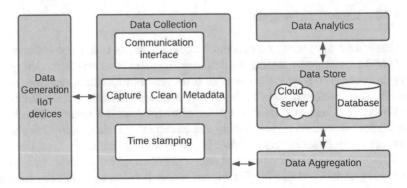

Fig. 13.6 Data flow at IDMS

13.6.3.4 Database Layer

The aggregated data is stored in local repositories to enhance the capability for computation and storage. This layer converts the heterogeneous industrial data into the structured format.

13.6.3.4.1 Database Management System

Database is a collection of related data which represents some feature of the real world. It is a structured group of records. A database system is designed to build and collate data for a particular application. **Database management system (DBMS)** is a software system for storing and retrieving user's data with specific properties with appropriate security measures. DBMS is the interface between the user and the database. DBMS accepts request for data from an application and instructs the operating system to provide the specific data. In large systems, DBMS allows users and third-party software to store and retrieve data. DBMS provides multiple user access and permits user to perform multiple operations like retrieve, update, and delete the records of choice. The structured data is organized into **four** types of databases: (1) hierarchical, (2) relational, (3) network, and (4) object-oriented database. The transactions on the database must follow **Atomicity, Consistency, Isolation, and Durability (ACID)** rules.

13.6.3.4.2 ACID Rules of DDBMS

Atomicity means the complete transaction which cannot be broken down to a set of similar transactions. When a transaction request is received by the database, the transaction has to be complete with no pending request. A transaction is an atomic unit of processing, that is, it is either performed in its entirety or not performed at all. No partial update should exist.

Data Consistency: Data after a transaction should remain consistent. A transaction should take the database from one consistent state to another consistent state and should not adversely affect any data item. That means if there are 100 records in the database and 2 of them are deleted with service requests, at the end of the transaction, there should be 98 records in the database.

Data Isolation: Each transaction has to be treated independently and has to be applied strictly on the record requested. There should not be any interference from the other concurrent transactions that are simultaneously running.

Durability: Durability means a completed transaction cannot be recalled and only a new transaction can be entertained on the database. If the committed transaction brings about a change, that change should be durable in the database and not lost in case of any failure.

13.6.3.4.3 Distributed Database (DDB)

Distributed database (DDB) is a collection of databases which are logically inter-related irrespective of where they are physically stored. They are location independent (meaning the user need not know where they are stored) but are logically related to each other. DDBMS is a software, which manages a distributed database system.

13.6.3.4.4 Consistency, Availability, and Partition (CAP) Tolerance

Consistency means every read of the database is guaranteed with the most recent record or a timeout error in the case of network failure.

Availability means every request receives a response with no guarantee that it is the information sought.

Partition tolerance means the system continues to operate despite arbitrary numbers of failed messages. Since network failure is inevitable, the partitioning needs to be tolerated.

13.6.3.4.5 Consistency, Availability, and Partition (CAP) Tolerance Theorem

CAP theorem as applied to DDBMS states that "it is not possible to guarantee all three properties (consistency, availability and Partition) simultaneously on a DDBMS." This is due to the network property which cannot be guaranteed all the time. There can be network failure at any point of time, and so there will be either availability of database access or consistency all the time.

13.6.3.4.6 Query Processing [1]

When a query is placed for fetching a data from the DDBMS, the query is scanned, parsed, and validated. The query is then internally represented by a query tree or a query graph. Alternate query search strategy is evaluated, and a suitable one is chosen to fetch the data from the database distributed in multiple locations. The resultant data is accumulated and returned. The major challenge in query processing is finding the most suitable strategy to fetch the data rather than the best strategy. This is because the query response time depends on delay in communication links, processing delay in distributed local subsystems, disparity in local resources for processing capabilities when its turn comes, and time to assemble resultant data from different locations. Query processing is shown in flowchart in Fig. 13.7.

13.6.3.4.7 Relational Algebra

Relational algebra plays an important role in query processing. The query after validation in the process flow shown in Fig. 13.7 is converted to relational algebraic expression before optimization.

Relational algebra defines the basic set of operations of relational database model. A sequence of relational algebra operations forms a relational algebra expression. The result of this expression represents the result of a database query.

The basic operations are:

- Projection
- Selection
- Union
- Intersection
- Minus
- Join

Projection

- **Projection** operation displays a subset of fields of a table. This gives a **vertical partition of the table** that can be used to get a part of the table, field-wise, or a section of the field with a criterion. If, for example, Student is the name of the student database shown in Table 13.1, with their details of class and course opted, projection will let one get the names of the students who have opted for a particular course.

Selection

Selection operation displays a subset of tuples of a table that satisfies certain conditions. This gives a **horizontal partition of the table**.

Fig. 13.7 Query processing flow

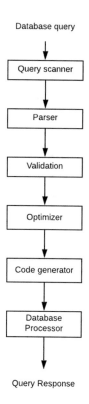

Database query

Query scanner

Parser

Validation

Optimizer

Code generator

Database Processor

Query Response

Table 13.1 Student database

Student				
Roll_No	Name	Course	Semester	Gender
2	Amit Tandon	BCA	1	Male
4	Prasad Biden	BCA	1	Female
5	Asif	MCA	2	Male
6	Joe Gorden	MCA	1	Male
8	Shaila Iyengar	BCA	1	Female

For example, in the Student table, if one wants to display the details of all students who have opted for MCA course, selection can be used.
• Usually, a combination of projection and selection operations is used to select the data from the database for processing.

Union

If P is the result of an operation and Q is the result of another operation, the union of P and Q is the set of all tuples that is either in P or in Q or in both without duplicates.

• For example, to display all students who are either in Semester 1 or are in BCA course, union is used.

Intersection

• If P is the result of an operation and Q is the result of another operation, the intersection of P is the set of all tuples that are in both P and Q.

Minus

• If P is the result of an operation and Q is the result of another operation, P - Q is the set of all tuples that are in P and not in Q.

Join

• Join operation combines related tuples of two different tables (results of queries) into a single table.

Branch

Branch is used to list fields along with their related information, converting SQL queries into Relational Algebra

A query is at first decomposed into smaller query blocks. These blocks are translated to equivalent relational algebra expressions. Optimization includes optimization of each block and then optimization of the query as a whole.

Few popular database structures are the following:

• MySQL
• Microsoft Access
• Oracle
• PostgreSQL
• dBase
• FoxPro
• SQLite
• IBM DB2
• LibreOffice Base
• MariaDB
• Microsoft SQL Server, etc.

13.6.4 Application and Management Layer

Application services and management functions are managed at this layer. It provides security and easy access to different data storage services. Industrial standard protocols are implemented for privacy and protection of the data and the end users of the system. The application layer provides a platform to extract useful patterns from industrial data and convert it into knowledge that is used for future improvement, early and better decision-making, and novel business opportunities. Information produced by industrial data is delivered as the set of services to the end users such as product lifecycle management (PLM), enterprise resource planning (ERP), supply chain management (SCM), manufacturing execution system (MES), quality management system (QMS), and warehouse management system (WMS). Various machine learning (ML) and deep learning (DL) techniques such as artificial neural network, random forest, support vector machine, logistic regression, recurrent neural network, restricted Boltzmann machine, autoencoder, and convolutional neural network can be applied on the industrial data for real-time and multitude of IIoT applications This layer also contains data analytics component. **Data analytics** is responsible for converting context-aware data into intelligent data. This intelligence is appreciated by analytics and can provide both delay-tolerant and delay-sensitive applications at the shop floor and middleware layer. The open cloud storage server is configured to support real-time queries and data coming from the IIoT middleware. The cloud computing techniques enable storage of massive data from factory devices and provide ubiquitous access to useful information for decision-making and collaboration among various industrial tools. The IIoT applications provide on-demand services through a cloud server which extends beyond the need to store sensor data in a scalable manner.

13.7 IoT Data Analytics

IoT data analytics provides analysis of large data (big data) generated by connected IoT devices. The data collected from IoT device is worthless if it is not analyzed. IoT data analytics are proven to give an organization the following advantages:

- Increased productivity
- Effective resource utilization
- Reduces leakages and helps optimization
- Evidence-based decision-making
- Planned maintenance by predictive analysis resulting in zero downtime
- Good supply chain management
- Enhanced customer experience

13.8   IoT Data Analytic Methods

The raw data collected and transmitted to cloud server contains a lot of irrelevant, redundant, repetitive information and unclear data. Few examples are: Blood pressure data captured using BP monitoring device containing multiple readings, Out of range readings due to wrong usage of devices, which are not the correct data set for assessing the health of the user; Addresses of employees of a company containing addresses of its clients and suppliers. It is therefore essential to clean the data to extract only relevant information from the incoming data from the IoT devices directly for the analysis. This is an important process in data science. Quality of data analysis depends on the quality of data used for analysis. Poor-quality data may even make the algorithms fail to generate correct meaningful outputs. On the other hand, high-quality data can cause a simple algorithm to give you outstanding results. There are many data cleaning techniques. **Data quality** is determined by determining the degree to which the data comply to the set of rules defined for a particular requirement. It is also called data validity. The errors in data validation occurs when data generating methods are not checked and validated for correctness. Multiple constraints are applied on data set to make the data valid. They are:

- Range: It is essential that some data has to be within the valid range. For example, human systolic and diastolic blood pressure range will be 80mmHg–180mmHg including abnormal condition. Any data which is not close to this range can easily be classified as errored information.
- Data type: The data has to be of a particular type, such as numeric, text, Boolean, date, etc. And there is no possibility of numbers (other than 0,1) in Boolean type of data set. Also, gender has to be male, female, or others, and it cannot be anything else.
- Compulsory constraints: These are mandatory restrictions on data. For example, patient identity number cannot be empty when there are vital data in the row of health data set.
- Unique requirement: Each data should have unique identification. For example, the social security number of two records cannot have the same social security numbers.
- Regular properties: Certain numbers or data fields have some specific properties, for example, phone numbers can be 10 digits in a particular country, in which case the field can be formatted to accept only 10 digit numbers for phone numbers when they belong to that particular country.

Data cleansing methods apply rules similar to the above to extract the correct set of data for analysis.

Clean data does not mean good-quality data. There are some more parameters to be considered to derive good-quality data for analysis. They are as follows:

Accuracy of data: Once the clean data is extracted, it is essential to check the data fields to be of right formats, types, etc. For example, the email address of the record may have additional characters, phone numbers which are valid may not be from the right region, etc.

Accuracy with precision: Data may be valid and accurate but may not be precise to the extent it is required. For example, the name may have the first name without the last name, and it is difficult to precisely identify as there can be many people with same names. It is either initials and/or surnames differentiate them to correctly identify. Another example is if the address is New York and there are no street and location, so there is lack of precision.

Completeness: Completeness is the degree to which you know the required values. Completeness is a little more challenging to achieve than accuracy or validity. Many times, the data field is filled with "not applicable (NA)" and "unknown (X)," for unclear data which still will be incomplete for analysis.

Consistency: Consistency is another important property determining the quality of a data set. It is achieved by comparing two or more related data fields in the data set. For example, if the **age is 15** and the other field is filled as "senior citizen," it is difficult to get the correctness of the data record, but by knowing whether the person is a senior citizen or nor, range can be set for the age field while cleaning the data.

Uniformity: The uniformity of the data may be in terms of unit of the parameter, for example, while collecting temperature as a parameter, it should be ensured that the instruction need to be given to feed in the data in Celsius or Fahrenheit. **temperature**, the unit is **Celsius or Fahrenheit**.

13.8.1 Data Cleansing Techniques

Data cleansing techniques are used to identify the errors and inconsistencies and missing values in the database. There can be one or more data cleansing techniques used to clean the data depending on the data set or analysis planned on the database. Some of the cleansing techniques used are:

- Removing irrelevant values
- Filling the missing records
- Converting the data to the common units
- Converting data types
- Collecting the correct data fields
- Reformatting all the data fields to the common formats of the data fields
- Removing redundant data records
- Avoiding types and errors

13.9 Analysis of Data

Depending on the application, type of analysis is chosen from, statistical, analytical and predictive to interpret data to arrive at conclusions [2]. Different analytical techniques as a function of their value generation are shown in Fig. 13.8. Different types of statistical analysis are the following:

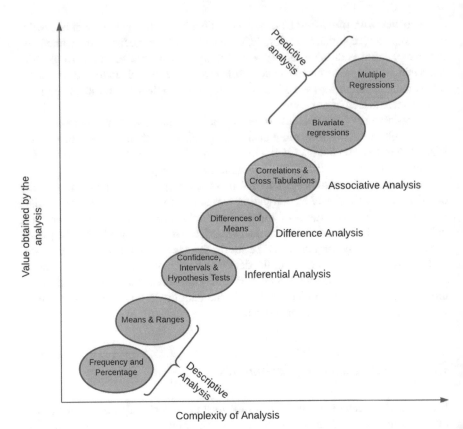

Fig. 13.8 Classification of analysis and their value addition

1. Descriptive analysis: This analysis describes the data set.
2. Inferential analysis: This data analysis helps to arrive at a conclusion.
3. Associative analysis: This data analysis determines the strength and direction of relationship between two or more data sets.
4. Differential analysis: Differential analysis compares the means of two or more similar data sets.
5. Predictive analysis: This analysis predicts future events based on the trend.

13.9.1   Predictive Analysis of the Data

Prediction of future events is done based on the analysis of historical data. Also, prediction is done depending on the application and the domain. Different types of predictive analysis are:

- Classification model
- Clustering model

- Forecast model
- Outlier model
- Time series model

The approach to predict from the data analysis can be broadly classified into two groups: **machine learning** and **deep learning**. Machine learning requires structured data with linear and non-linear data sets. The machines will be trained by the data set and eventually made to look for similar events in the future. Deep learning is a subset of machine learning which deals with audio, video, text, and images. The deep learning algorithms process data in layers, form artificial neural network, and learn and predict using intelligence.

13.9.2 Data Visualization

Data visualization is the graphical representation of information and **data**. By using visual elements like charts, graphs, and maps, **data visualization** tools provide an accessible way to see and understand trends, outliers, and patterns in **data**. There are visualization tools which will convert the database to visual form for easy human consumption and interpretation. Some of the most used visualization tools are Tableau, Looker, Zoho Analytics, Sisense, IBM Cognos Analytics, Qlik Sense and Domo, and Microsoft Power Bi.

References

1. www.tutorialspoint.com
2. https://www.slideshare.net/gondaliyamehul/data-analysis-and-interpretation-43140630
3. Muhammad Saqlain, Minghao Piao, Youngbok Shim, Jong Yun Lee, Chungbuk National University, Cheongju, "Framework of an IoT-based Industrial Data Management for Smart Manufacturing", Journal of Sensor and actuator, MDPI, 28 April 2019

Chapter 14
Design for Manufacturing and Business Models

14.1 Introduction to Design for Manufacturing (DFM)

The product development has to lead to marketability and commercial success. This can happen when the product is affordable (positioned at the right price which the customer can pay). The product will be affordable if positioned at low cost for which its necessary to maintain the design and development cost which is called non-recurring engineering (NRE) cost at its minimum. The NRE cost is amortized with the profit margin over the years. This must be considered from the design planning stage of the product development. IoT devices are no exceptions to this, and they are more sensitive to the market dynamics. Hence, it is doubly important to keep in mind the factors affecting product cost to keep it at minimum. There are various aspects, which contribute to the product cost. They are:

1. The specialization of labor including designers who can develop.
2. Use of integrated technology or automation in production.
3. Cost of product on the market (affordability of the user and demand in the market). Lower per-unit cost can come from bulk orders, larger advertising buys, or lower cost of capital. (Cost of capital is the required return to justify the investment of product development and commercialization.)

14.1.1 Economies of Scale

Economies of scale are cost advantages companies get when production becomes efficient, as costs can be spread over a larger number of products. It can be maximized both by controlling both internal and external parameters. Economies of scale can be achieved internally by optimizing the operational and infrastructure costs during design and development. This has limitation of more time to market.

It depends on factors outside the company which are considered external in the same industry. They are mainly the new production methods, logistics of production and assembly, customization needs, region of market, market demand, government policies, lower production and operational costs, etc. Individual firms have no direct control over what happens externally. Only thing which is internal to companies is to reduce the production or manufacturing cost by design. This is called *design for manufacturing* (DFM).

14.1.2   Design for Manufacturing (DFM)

DFM is the design methodology which aims to lower the cost of production by optimizing the design workflow, production automation to target ease of manufacturing and assembly. In simple words, DFM aims to make products at low cost with reduced time to market. The acronym Design for Manufacturing and Assembly (DFMA) is sometimes termed as DFM. Device making involves design, process, materials, environment, and compliance and safety. It is important to keep in mind the manufacturability and material suitability early in the design stage to avoid extra cost for iterations later during the development stage. The considerations are biocompatibility in the case of human wearable IoT devices, safety (for both environment and people), interoperability (e.g., the OCF device to oneM2M communication needs), and interference (co-existence with other devices which are operating in the same place especially for radio frequencies of Bluetooth, Wi-Fi devices without interfering, etc.) Tooling requirements of mechanical parts for the IoT devices need to be considered early in the development stage. The importance of DFM is required to be known to all the stakeholders—engineers, designers, contract manufacturer, mold developer, and material supplier. The intent of this "cross-functional" DFM is to challenge the design—to look at the design at all levels, component, subsystem, system, and holistic levels, to ensure the design is optimized and does not have unnecessary cost embedded in it. It is to be noted that the cost of the development would be 10% more for correcting issues detected at one stage later than the previous (impact of change). So, it is absolutely essential to consider every aspect of it early in the design stage. It is very essential to follow the five principles early in the design stage to achieve DFM.

14.1.2.1   Design

It is required to follow proper design guidelines and processes as relevant for the product/device. In the case of mechanical design, it is very essential to have complete understanding of the product assembly and fitment with the electronics department so that the design works as intended in the first attempt. All the stakeholders should have clear understanding of the designs.

14.1.2.2 Process

Certain IoT products or devices have to be designed and developed using standard processes defined by the standardization bodies. For example, the design process for medical technology devices should comply to ISO 13485 for the product to be sold in the market. It has to be ensured at the beginning of the design cycle that these processes are well understood and followed during the process. It is better to involve one control person to monitor these requirements during design, development, and production processes.

14.1.2.3 Material

It is very important to use good quality material to maintain the minimum acceptable quality of the devices. Also, there requirement to comply with quality standards to be followed to get this right. For example, the devices using chargers for certain applications have to use UL-certified batteries as per the regional regulatory and safety needs. Such requirements have to be addressed early in the design stage to avoid any repetitions. Other applicable requirements to be cosidered early in the design stage to avoid redesigns. An example for this is that the bio medical devices which are wearables are to be made using bio compatible materials.

14.1.2.4 Environment/Testing

The device must be designed to withstand the dynamically varying environmental conditions in which it is expected to work. It is guaranteed by testing for electromagnetic compatibility (EMC) so that it works as expected to function satisfactorily in an electromagnetic environment without introducing intolerable disturbances to the environment. To ensure this, the devices have to comply with environmental standards. These tests include electromagnetic susceptibility (EMS), emission (EME), and immunity (EMI). For example, IEC 60601-2-2-2017 standard defines the requirements for the basic safety and essential performance of high-frequency surgical equipment and accessories. Similarly, there are standards, which define the minimum performance requirements of devices under different environmental conditions. There are test houses with facilities which can test devices by creating these dynamic conditions environments and test the impact and effect on the functionality of the devices. They also certify these devices for the tests conducted by them against the relevant standards.

14.1.2.5   Regulatory Requirements

In addition to the intended functionality of the product and the environmental requirements, it is necessary to understand the regulatory standards prevailing in the region of the targeted market. The region-specific regulatory bodies have defined local standards on the product specifications. It is mandatory to comply with the regulatory requirements to market a product in the region. These specifications have to be considered at the design stage to make the product marketable in the target region.

14.2   Factors Affecting DFM

Major factors to be considered for the DFM during early design stage for the devices are the following:

14.2.1   Part Count

It is essential to minimize the part or component count which goes into a product design. The smaller the number, the lesser the complexity in product assembly. It is better to understand and classify parts, which will be automatically assembled. This helps plan part placement on the product providing enough space for manual assembly. It is all the more essential if the parts are being sourced from vendors who are from different places so that the risk of out-of-stock scenario can be mitigated.

14.2.2   Standard Parts and Materials

It is better to use standard parts and easily available materials for minimizing the cost of manufacturing and avoiding complex logistics of material sourcing. When it comes to electronics components and modules, it is necessary to look at the obsolescence and continued service support. Electronics modules go out of stock very easily, and hence it is a regular practice to have two or more alternatives' parts for most essential modules of the devices. Also, one needs to check the availability of the parts in reels which can be directly mounted onto electronic assembly lines for large-scale production.

14.2.3 Modularity

It is essential to consider the ease of long-term customer support. It is a good practice to make the design modular so that whenever a module is found faulty, it can be isolated and easily replaced without much functionality changes.

14.2.4 Design for Assembly

Product assembly is another major consideration in manufacturing the devices in large number. Hence, it should be kept in mind that assembly of parts is a major time-consuming phase in production. It is necessary to design the device such that it can be easily assembled, checking for interlocking of parts, easily fitment of the screw positions, alignment of screws and bolts of display. Keypad modules are to be selected so that their assembly is easy. As much as possible, design should be such that there is no or minimum manual interaction during production and assembly.

14.2.5 Aesthetics and Surface Finish

It is always important to have good aesthetics and great surface finishes for the products. Good aesthetics is very subjective, and a combination of aesthetics and comfort in use are to be considered during the product design. This requires a lot of processes and, hence, time. Good practice, therefore, is to define aesthetics and surface finishes acceptable and optimize the cost of production and shorter time to market.

14.3 Product Documentation and Media Materials

It is essential to have good product documentation for easy use of devices. It is to be remembered that devices are used by the people who have no prior knowledge of how it is to be used. It is essential to consider stakeholders for the product like marketing, sales team, and customers. The document is the way to demonstrate the use and advantages of the device. Nowadays, "how to use" videos demonstrating the functionality of the product is a great way to demonstrate and educate the usage of the product. In most times, post-sales support gets simplified enormously with proper product support materials.

Good product document should include **screenshots and pictures** of the product in different working scenarios. Many prefer to go through images than reading line-by-line instructions. **Examples** are the best way to explain the usage scenarios of the product. **Document updates** are to be transparent to users as they are the intended users.

Main product documentations include:

- Installation procedure
- Application notes
- Usage scenarios
- Information regarding product certifications
- Link to videos and audio files explaining product usage from unboxing to usage scenarios
- Customer reviews

14.4   Business Model and Business Model Canvas

Business model is a global concept which defines how a business can create, capture, and deliver value for itself while offering its products or services to its customers. It is often mistaken to be a revenue model, production model, or sales model, but it is a big picture which includes all of these. It is easily represented by the nine-component canvas as shown in Fig. 14.1.

The nine components of the business model are the following.

14.4.1   Value Proposition

The most important in the business canvas is the value proposition. Any business should have a clear answer to what problem is it solving or which need is getting fulfilled by the product or service offered by the company.

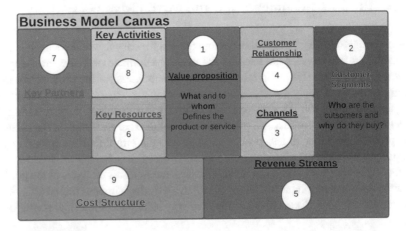

Fig. 14.1 Business model canvas

14.4.2   Customer Segment

Who is the target customer? The answer to this question will define the target customers and the target market. Also, answer to why do they buy the product and service offered and why from this company will help in defining the customer-specific solution and unique selling proposition offered by the company. This defines geographic boundary of the target market for the business.

14.4.3   Channels

Sales and distribution channels help logistics of providing products and services to the target customers. This can be physical channel of distribution or virtual channels on digital platforms for services.

14.4.4   Customer Relationship

Customer relationship helps define strategies for getting new customers and retaining the existing customers and growing them. It is related to channels, customer segment, and the value offerings to the customers defined above.

14.4.5   Revenue Streams

What is the value of the product or service that the customer will pay for? The answer defines the revenue generation model or revenue model. The types of revenue models are subscription or direct sales of product or service. This has to come by interacting with a large number of customers or prospective customers. Arriving at the correct revenue model for the product or service a company deals with is very challenging.

14.4.6   Key Resources

Key resources required have to be defined based on how a company wants to develop and sell the product or services to the target customer. The main resources are finance, physical assets like design tools, infrastructure, and manufacturing setups, raw materials; and intellectual properties like design know-how and human resources to execute these activities.

14.4.7 Key Partners

Who are the partners needed to come together and what type of relationship needs to be built with the identified partners to make this happen? This is done by defining partners and their roles in the company. They are suppliers, employers, manufacturing partners, financial agencies, and marketing partners. This can lead to a large risk if not managed well. Identifying the right resources in the right numbers and the type of relationship is crucial to the success of the company.

14.4.8 Key Activities

Key activities are needed to make the business model work. The key activities could be design, production, logistics, services, etc. which are important for the company.

14.4.9 Cost

This is the major component in business which defines costs incurred to make the business model work. This is another challenging activity where companies fail if not handled properly. To manage costs well, it is necessary to know which of the activities incur more cost, which are expensive resources, etc. to optimize them. This further leads to defining fixed costs, variable costs, economies of scale, operating cost, etc.

14.5 IoT System Development

IoT system development is a detailed development activity which should consider the time-to-market duration and the features required in the system solution. It involves development of the proof of concept, prototyping and validation, and winning a first paying customer.

14.6 Business Models for IoT Solutions

Businesses involving products and services around IoT and M2M communication are different. They also need to manage all the above components of business model canvas well to succeed. It could be the simplest activity like defining positioning strategy of the product, right campaign, and marketing it as per the company positioning. If the company already has products in the market, it is better to adopt incremental strategies like adding the connectivity to the Internet and capture customer behavior and preferences. This gives a lot of information which can be used for fine-tuning the marketing strategy in the business. It is very essential to decide whether the product has to be positioned as consumer product (B2C) or for wholesale business entity (B2B) which is a major decision as the factors which govern the sale are different. The product selling cost in B2C market is higher than the cost in the B2B market. Services around IoT system depend on the quality of service-based business. QOS is the quality of deliveries which could be service or product and ability to provide service on the call.

Another important consideration of business model is deciding on how the IoT solutions are marketed. There are various models one can explore depending on the market dynamics and the solutions offered. They are:

1. Subscription services around IoTs
2. Outcome-based models
3. Asset sharing models
4. Things as a service

14.6.1   Subscription Services Around IoT

As it can be seen, IoT based solution can be deployed in almost all domains. It is very essential to maximize the profitability in business for sustainability. Selling IoT solution is a one-time sale opportunity, but offering services around them would be a recurring revenue option which has to be considered. The subscription model around IoT solution allows business to generate recurring revenue. In other words, in addition to one-time sale of the product, value added services around the product will assure recurring revenue to the companies.

14.6.2   Outcome-Based Models

There are companies which setup high cost infrastructures like cloud servers, and high cost tools on cloud servers which are offered for use by the customers and are charged as they use. This means investment for the IoT infrastructure like servers, cloud subscriptions, devices, and network entities will be setup by the companies, but another set of service companies will operate on them to generate the desired results for the company which owns the infrastructure. This typically works for logistics management and administrative automation kind of activities using IoT solutions.

14.6.3 Asset Sharing Models

The model allows businesses to share their costly **IoT-enabled assets** with other business entities. The organization sharing the asset can charge based on use, duration of use, and nature of usage.

14.6.4 Things as a Service

An IoT-as-a-service model is the most commonly used business model today. In this model, a business provides its IoT-enabled solution on lease and generates revenue. The model is not limited exclusively to software or physical products and "things", but also revenue is generated from selling information and data; The *IoT-as-a-service model* can be considered as a part of the abovementioned subscription model and can generate recurring revenue. This is also extensively used for providing cloud-based data maintenance services.

14.7 Commercialization

Commercialization of IoT devices and services poses different challenge compared to engineering. Commercialization challenges are:

1. Access to large scale manufacturing capability (not applicable for IoT services)
2. Access to marketing and distribution channels
3. Sales and customer support
4. Cost of the product/services
5. Promotional campaigns
6. Continued support and services around the post-sale issues by taking the issues to engineering and getting it resolved
7. New product initiation depending on the customer requirement leading to product life cycle
8. Supply chain logistics management
9. Maintenance support and services
10. Brand image building and retaining it

It is important to understand these activities and the challenges involved in them before one goes for commercialization of the IoT systems. Each of them are discussed in following sections:

14.7.1 Manufacturing Capability (Not Applicable for IoT Services)

Once the IoT system prototype is ready and prospects are lined up, it is essential to manufacture IoT devices in large numbers to get advantage of economy of scale. Making IoT in small numbers will cost at least 200% more. To manufacture in large numbers, one should have market reach and large target addressable market segment and other market resources needed to address it. This is the chicken and egg situation which is a major challenge at least when it comes to IoT devices. It can only be compensated by multiple marketing strategies discussed above.

14.7.2 Marketing Channels

Marketing and distribution channels are very essential to address the identified market segment. This determines the business success along with the right market strategy. Marketing channel consists of marketing personnel, distributors, marketing representatives, advertisement resources, and the sales teams distributed across different regions of the target market. Based on the market strategy, optimization of marketing cost by managing the marketing channel resources helps get commercial success for the product.

14.7.3 Sales and Customer Support

After-sales and customer support play a crucial role in the success of the product. The after-sales customer support activity is to make the customer comfortable in using the product, get the best of his experience, and influence others to buy them.

14.7.4 Cost of the Product/Services

Cost of the product or service has to be appropriately placed at the product launch so that the prospective customer will see higher value than the money they pay for the product. After placing the product in the market, it is essential to continue to optimize the cost of the product development to maximize profit margin during the product life cycle. This activity concentrates on optimizing the cost of production logistics or service logistics.

14.7.5   *Promotional Campaigns*

Market capture and new market development is greatly influenced by the product campaigns which is typically a marketing activity. This has greater relevance in current days than in the past with access to the digital marketing.

14.7.6   *New Product Development*

It is obvious that every product has a product life cycle. It is necessary to track the product life cycle curve and initiate a new version of the product with newer features depending on the customer feedbacks which will ensure continuity in business for the product lines.

Type of Business Models

IoT and IIoT devices and communication provide a great opportunity with a bouquet of business models. Conventional business models depending on the parties participating in business transactions are the following:

- Business to business (B2B)
- Business to consumer (B2C)
- Consumer to business (C2B)
- Consumer to consumer (C2C)
- Business to government (B2G)

Since IoT system solutions are multiple domain solution involving skill sets spread across domains, products, services, maintenance, and distributed infrastructure, one can imagine business opportunities in all the abovementioned models.

The business activity with IoT system solutions for Industry 4.0 is a three-stage process as shown in Fig. 14.2.

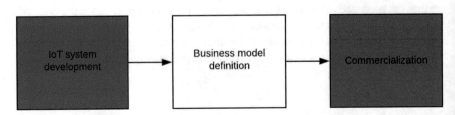

Fig. 14.2 Business flow

It involves developing the IoT-based system solution for the identified problem, developing the business plan for positioning it in the market, and commercializing it. Each of this is an involved process as explained in brief below.

14.8   Business Model Innovation

Increased pace of innovation and global reach through digital media and other resources, it is obvious that innovation has paved way to business model. Business innovation is the only tool to do business and maximize the value to all stakeholders. Recent flexible subscription models, Freemiums, platforms as service, and direct to customer are the result of such business model innovation.

The pandemic during 2019–2021 encouraged the adoption of subscription models of services like medical consultancies, COVID packages (involving medical consultation, delivery of medicines, and logistics of delivering essentials), etc. **Freemium** is the new term coined by combining free and premium value which means that the products and services are offered free of cost initially for some time or for limited-order quantity beyond which a premium price is charged. This model is familiar for applications such as Zoom conferencing platform, Spotify, iCloud, etc.

Direct-to-customer model is quite familiar with business support on messengers like WhatsApp, Facebook, etc. apart from the regular online shopping platforms. All the above examples are business marketing innovations, but there are innovations happening in the production, supply chain logistics, and other components of the business model as well.

The business model innovation offers advantages like maximizing value, increased revenue by reduced overheads, competitive advantage, and brand recognition. It is a continuous process, and the sooner companies adopt it, the longer can success be sustained.

Chapter 15
Question Bank and Reference Design Material

15.1 Part 1

15.1.1 Chapter 1

1. What is the relevance of Industry 4.0?
2. Do you know the history of Industry revolution?
3. Explain the role of IoT in fourth-generation Industry revolution I4.0.
4. Explain the IoT concept.
5. List and explain the characteristics of IoTs.
6. What is Industrial Internet of Things (IIoT)?
7. Explain the different technologies applicable to IoT.
8. Explain with diagram the layer architecture of IoT.
9. Compare the layer architecture of IoT and OSI communication models.
10. What are the features of the ETSI high-level architecture of IoT?
11. What are active and passive IoT devices?
12. What is M2M communication?

15.1.2 Chapter 2

1. Explain the architectural framework of IoT system solution.
2. Explain the three-layer framework for IoT system.
3. Explain the five-layer framework of IoT system.
4. What is edge computing architecture for IoT solution?
5. Describe the roles of cloud, fog, and edge in IoT architecture.
6. Design conceptual IoT system for a home automation using standard IoT design flow. Explain the steps.

© The Author(s), under exclusive license to Springer Nature Switzerland AG 2021
V. S. Chakravarthi, *Internet of Things and M2M Communication Technologies*,
https://doi.org/10.1007/978-3-030-79272-5_15

7. How do you identify the requirement for a problem and convert it to technical specification?
8. How do you identify the layer architecture for the proposed IoT system?
9. How do you choose the relevant technologies for an IoT problem?
10. What is the need to comply to standards for IoT system? Name some applicable standards you know.
11. Where do you use firmware in IoT system? Write a note on firmware development environment.
12. What is the role of communication technology in IoT systems?

15.1.3   Chapter 3

1. How is Industrial IoT different from IoT?
2. What are the challenges of development of IIoT and how are they different from IoT?
3. Explain the automation framework using IIoT.
4. Write a short note on:

   (a) Single-board computer
   (b) Data server
   (c) HMI and supervisory systems
   (d) SCADA software

5. Explain sensors and actuators needed for IIoT-based factory automation system.
6. What do you understand by distributed ledger technologies (DLTs)?
7. Explain any two IIoT development platforms of your choice.
8. What are the different IIoT subsystems?
9. Explain the relevance of communication technology in IIoT system.
10. How is IIoT development different from IoT development?

15.1.4   Chapter 4

1. Plan the development of IoT-based environment monitoring system.
2. What are the product requirements of your IoT-based environment monitoring system?
3. What is proof-of-concept system? Why is it required for product development?
4. Design a proof-of-concept system based on IoT for environment monitoring.
5. How do you decide on the component/modules required for the product to be developed?
6. How do you define the software requirement for the IoT product?
7. What is prototype and how does it help in IoT product development?

8. What is the relevance of aesthetics of a product and why are they important?
9. How do you decide on cloud services needed for IoT system?
10. What are the factors considered in deciding on the cloud platform from a wide choice of cloud platforms available?

15.1.5 Chapter 5

1. What are sensors and actuators? How are they used in IoT and IIoT systems?
2. What are the different types of sensors?
3. Why do you need signal conditioning for the outputs of a sensor?
4. Explain the functions of signal conditioning circuits.
5. Where do you need electrical isolation in the sensor devices?
6. Explain the role of microcontroller in sensor device.
7. What are the main features of NVM protocol-based storage memories for IoT devices?
8. What are hardware accelerators? How are they used in IoT context?
9. With block diagram, explain the function of hardware accelerators.
10. Differentiate NVM and SSD storage in IoT and IIoT context.

15.1.6 Chapter 6

1. Identify applications which need capturing modules in the IoT systems.
2. Name a few capturing modules and how they are used in IoT and IIoT solutions.
3. Define the requirement for IIoT-based system for monitoring working personnel in the factory premises. What parameters would you wish to monitor in these scenarios?
4. Explain the internal blocks of CCD camera module.
5. What are the different types of camera modules which find use in IoT systems?
6. What are the factors you consider while identifying the right type of capturing module for your IoT system?
7. Differentiate CCD and CMOS cameras and define how would you use either of them.
8. Write a technical note on smartphone camera modules with detailed specifications.
9. How are videos captured and how are they used in IoT applications?
10. How do you interface camera module in IoT system?

15.1.7   Chapter 7

1. Explain the different layers of software required for IoT system with their functionalities.
2. With the block diagram, explain IoT device architecture with relevant software protocols applicable in IoT product.
3. How do you select the hardware requirement for an IoT product development?
4. Explain the method to define requirements of IoT device software.
5. What is firmware upgrade? How is it done remotely? What are the advantages of the same?
6. What is IoT security? What is the role of IoT software in it?
7. What is IoT embedded system software?
8. Explain the features of embedded software development platform.
9. What are the options for developing embedded software for IoT system?
10. What are the typical functions of IoT firmware?
11. What are the computational and communication functions in IoT systems?
12. Explain the embedded software architecture guidelines you would follow in software development for IoT system.
13. How do you embed the software in hardware? What is this process called?
14. List the most used IDEs and compare their features.
15. What is cloud computing? How is it relevant in IoT context?
16. What are edge development platforms and when do you use them?

15.1.8   Chapter 8

1. What is the need for IoT security?
2. What are the vulnerabilities of IoT network?
3. What are the challenges of IoT system which call for IoT security?
4. Define and differentiate (1) privacy, (2) security, (3) interoperability, and (4) data integrity.
5. What are the components in IoT system where end-to-end security is considered? Explain.
6. What are the essential cryptographic algorithms used in IoT security?
7. Explain IoT security platforms and functions of different subsystems.
8. What are the different IoT security cloud services offered?
9. Explain how security is implemented using layer-wise IP stack functionality.
10. What are cryptography keys and what are their importance in IoT security?
11. Explain IoT security tomography and layer attacker model.
12. List the best practices to be followed to ensure IoT security and explain how they help in maintaining security.

15.1.9   Chapter 9

1. What is the relevance of application layer in IoT context?
2. What are the different types of applications which are used in IoT system?
3. Explain CoAP IoT network and the operations.
4. How is IoT application different from regular application?
5. Explain CoRE and CoAP protocol in IoT network.
6. What is MQTT and explain the protocol.
7. When is MQTT protocol used in IoT networks?
8. What is XMPP? How does it operate in client-server configuration?
9. What is WebSocket and how is it helpful in IoT system?
10. Explain SOAP protocol.
11. Explain the main features of REST and RESTful APIs.
12. How are RESTful APIs different from HTTP APIs?
13. Explain in brief the application development platforms. How are they being used to develop cross-protocol applications?
14. List a few commercial application development platforms and compare their features.
15. Describe a case study of IoT system application which uses all these protocols.

15.1.10   Chapter 10

1. What is M2M communication? Where do you need it?
2. What are the two main classifications of M2M communication? Give examples of each of them.
3. Explain the different communication technologies used for M2M communication.
4. Why do you think wireless technologies are most used in M2M communication?
5. Which are short-range communication technologies deployed in M2M communication?
6. Explain in brief the history of evolution of standards and protocols for M2M communication.
7. Compare and contrast different technologies in terms of their specifications which can be used for M2M communication.
8. Explain the architectural framework of M2M communication.
9. Describe oneM2M functional architecture for IP-based M2M communication.
10. Describe oneM2M functional architecture for non-IP-based M2M communication.
11. Describe your understanding of how communication happens between peer-to-peer layer-wise M2M communication model.
12. With the diagram, explain the conceptual model of M2M device.
13. Explain the IoT-based applications where M2M communication can be deployed.

14. What are the different application protocols used in M2M communication and can a device support multiple protocols?
15. With the help of application scenario and a corresponding message sequence chart, explain how M2M communication happens.
16. Explain the functioning of AI-enabled home light control system with the help of message sequence charts.
17. Describe the highlights of the following protocols:

    (a) HTTP
    (b) CoAP
    (c) MQTT

18. Justify the use of various technologies for the different scenarios which are used for M2M communication.

15.1.11   Chapter 11

1. Explain M2M communication with respect to IP protocol stack.
2. Describe your understanding of M2M communication using OSI layer architecture.
3. With a diagram, explain Internet protocol (IP) stack.
4. Explain IPv6 and its advantages in IoT systems.
5. Explain how M2M communication is achieved in the context of IoT to access the Internet.
6. Explain the contexts where TCP protocol is used for IoT M2M communication and how?
7. Explain the contexts where UDP protocol is used for IoT M2M communication and how?
8. Explain how the following protocols are used in IoT context:

    (a) Client-server configuration using HTTP protocol
    (b) Internet access in constrained environment using CoAP protocol
    (c) Subscribe-publish protocol in MQTT applications

9. Write a note on applicable standard defining bodies for M2M communication and their roles and responsibilities.
10. What is the Open Connectivity Foundation (OCF) and the communication mechanism defined by them?

15.1.12   Chapter 12

1. What are constrained devices and networks? Where are they used? Give examples.
2. How do these devices access the Internet?

3. Explain M2M communication and PAN.
4. Explain constrained device architecture. What are its main constituents?
5. What is wireless sensor network (WSN)? What is its relevance in the context of IoT?
6. Explain the different technologies deployed in WSN.
7. Explain the architecture of MBAN system.
8. Describe the network architecture of WPAN with the layered architecture.
9. Explain CoRE architecture for resource-oriented applications.
10. Write technical notes on the following protocols:

    (a) BT-LE
    (b) 6LoWPAN
    (c) DECT Ultra Low Energy
    (d) RoLL
    (e) Z-Wave
    (f) Power Line Communication (PLC)
    (g) NFC
    (h) BACNET
    (i) LPWAN
    (j) LoRaWAN
    (k) Sigfox

11. Explain the design consideration of IIoT systems for industrial automation.
12. How are industry automation classified?
13. Explain automation pyramid of industry automation.
14. Differentiate the design consideration of IIoT system with IoT systems.
15. What are the critical design considerations which are important for industry automation?

15.1.13   Chapter 13

1. Explain IoT database and analytics framework.
2. Describe big data life cycle for IoT network.
3. Describe IoT data management framework as mapped to big data life cycle.
4. Explain data framework for WPAN.
5. Describe big data life cycle for IIoT network.
6. Describe IIoT data management system (IDMS) framework as mapped to big data life cycle.
7. What are the ACID rules of DDBMS?
8. How is distributed database different from DBMS?
9. What is CAP tolerance theorem as applied to DDB?
10. Explain the different data analytic methods on the IoT database. Also, discuss how do you decide on what analytics to be run on the database.
11. Justify why data received from IoTs has to be cleaned?
12. What are the cleansing methods for database?
13. What are data visualization and different data visualization methods?

15.1.14   Chapter 14

1. Define design for manufacturing (DFM) in the context of IoT and IIoTs.
2. What is economy of scale?
3. What are the DFM considerations for the IoT devices?
4. What are the factors affecting DFM?
5. Explain the different business models and how to decide the suitable business model.
6. What are the considerations for commercialization of the solution?

15.2   Part 2

15.2.1   Prerequisites

The user should have working knowledge of scripts and languages like Embedded C, C++, Java or Python, and HTML which can otherwise be referred from different online sources. It is essential to have the development boards on which the firmware developed can be downloaded and run. It is good to refer to the official sites of standard bodies like IETF, MQTT, and XMPP for latest information.

Development Platform:

- Arduino boards
- Intel Galileo boards
- Silicon Labs Gecko boards
- Embedded software IDEs
- Mbed IDE
- Arduino IDE
- Studio
- Brief explanation of setup

This experiment illustrates the communication of serial data from the laptop where IDE is running which can be received back onto the laptop terminal. The setup is shown in Fig. 15.1. The UART on the development board is configured for 1.1 Mbits data rate. Laptop runs the IDE and terminal program to feed in the data to be transmitted. On loopback, the same data is echoed back through the receive lines of serial data which get printed on the terminal for the user to see. Arduino board has a built-in support for serial communication, UART on pins 0 and 1.

15.2.2   Experiment 1

Program to interface Arduino board serial port and loopback testing through UART
    Firmware code illustrates loopback the data through the serial port on the Arduino
board. As we input the data from the transmit terminal, the output is echoed back to
the same terminal to see.

```
void setup()
{
  Serial.begin (115200);                 // Open serial communications
  while (!Serial)
  {
    ;                                     // wait for serial port to connect.
  }

  Serial.println("Serial Communication");
}

void loop()
{
  if (Serial.available())
    Serial.write(Serial.read());          // reads and writes the data
on the terminal
}
```

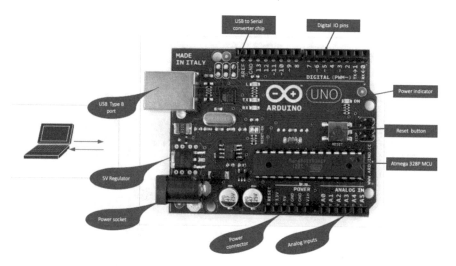

Fig. 15.1 Loopback test through the serial port on Arduino board

15.2.3   Experiment 2

Program to interface with ADC to Arduino board

Sensors and actuators are interfaced with the microcontroller through analog-to-digital converters (ADC) and digital-to-analog converters (DAC). The input outputs of Atmega micro controller are brought out at digital interface, and the analog interfaces are connected to the external transducers/sensors, and the relay actuators are controlled by analog output of DACs.

Brief explanation of setup is as follows:

The output of a sensor many times is analog in nature. The microcontroller processes digital signals. ADC connected in between a sensor and a microcontroller converts a continuous-time and continuous-amplitude analog signal to a discrete-time and discrete-amplitude digital signal which the microcontroller processes further.

Arduino Uno has six on-board ADC channels which can be used to read analog signal in the range 0–5 V. It has 10-bit ADC which means it will give digital value in the range of 0–1023 (2^{10}). This is called the resolution which indicates the number of discrete values it can produce corresponding to voltage between 0 and 5 V.

$$\text{analog voltage measured} = \left(\text{digital code} \times \text{Vref}\right) / \left(2^{n} - 1\right)$$

where:

Vref—The reference voltage, which is the maximum value that the ADC can convert

n—Resolution of ADC

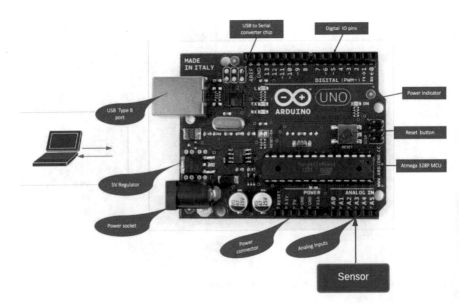

Fig. 15.2 ADC interface and reading the sensor value through the serial port on the laptop

The experimental setup is shown in Fig. 15.2. In the experimental setup shown, the sensor input is connected to A0 analog input of Arduino board, and the value is converted digital signal in the on-board ADC which is read through the serial port connected to the laptop. The value read is displayed on the terminal running in the laptop. The serial port is configured for 1,15Mbps data rate.

Firmware code

```
/* Setup the serial port to display the result on the terminal */
void setup()
{
  Serial.begin(115200);
}


/*  Monitor the sensor continuously with delay of 1000ms    */
void loop()
{
  int sensorPin = A0;                                    //
select the analog input pin to connect the sensor
  int digitalValue = 0;                                  //
variable to store the value coming from the sensor
  float analogVoltage = 0.00;

  digitalValue = analogRead(sensorPin);                       // read
the value from the analog channel
  Serial.print("digital value = ");
  Serial.print(digitalValue);                                //
print digital value on the terminal

  analogVoltage = (digitalValue * 5.00) / 1023.00;      // Vref=
5V,10 bit ADC (2^10)-1 = 1023
  Serial.print("  analog voltage = ");
  Serial.println(analogVoltage);                  //print voltage on the
terminal
  delay(1000);
}
```

15.2.4   Experiment 3

Program to interface with serial port through SPI

Brief explanation of setup

SPI uses the following four wires:

- SCK (pin 13)—The clock pulses which synchronize data transmission which is generated by the master.
- MOSI (pin 11)—The master sends commands/data to peripherals.

- MISO (pin 12)—The master receives data from the peripheral device.
- SS (pin 10)—Select line to select the peripheral to communicate.

The firmware code illustrates interfacing EEPROM to Arduino and a functionality to write and read the data from the module.

Connections:

Arduino Uno	AT25HP512
MOSI	SI
MISO	SO
SCLK	SCK
SS	CS
3v	Vcc, WP, HOLD
GND	GND

Firmware code

```
#define DATAOUT        11       //MOSI
#define DATAIN         12       //MISO
#define SPICLOCK       13       //SCK
#define SLAVESELECT    10       //SS

//opcodes
#define WREN           6
#define WRDI           4
#define RDSR           5
#define WRSR           1
#define READ           3
#define WRITE          2

byte eeprom_output_data = 0;
byte eeprom_input_data = 0;
byte clr;
int address = 0;
char buffer [128];

void fill_buffer()
{
    for (int i=0;i<128;i++)
    {
      buffer[i]=i;
    }
}

char spi_transfer(volatile char data)
{
```

```
    SPDR = data;                        // Start the transmission
    while (!(SPSR & (1<<SPIF))); // Wait the end of the transmission

    return SPDR;                        // return the received byte
}

void setup()
{
    Serial.begin(115200);

    pinMode(DATAOUT, OUTPUT);
    pinMode(DATAIN, INPUT);
    pinMode(SPICLOCK,OUTPUT);
    pinMode(SLAVESELECT,OUTPUT);
    digitalWrite(SLAVESELECT,HIGH);              //disable device

        // SPCR = 01010000, interrupt disabled,spi enabled,msb
1st,master,clk low when idle,
    //sample on leading edge of clk,system clock/4 rate (fastest)

    SPCR = (1<<SPE)|(1<<MSTR);
    clr=SPSR;
    clr=SPDR;
    delay(10);

    fill_buffer();                      //fill buffer with data


    digitalWrite(SLAVESELECT,LOW);
    spi_transfer(WREN); //write enable
    digitalWrite(SLAVESELECT,HIGH);
    delay(10);

    digitalWrite(SLAVESELECT,LOW);
    spi_transfer(WRITE);                        //write instruction

    address = 0;
    spi_transfer((char)(address>>8));   //send MSByte address first
    spi_transfer((char)(address));          //send LSByte address
```

```
    for (int i=0;i<128;i++)                    //write 128 bytes
    {
          spi_transfer(buffer[i]);              //write data byte
    }

    digitalWrite(SLAVESELECT,HIGH);            //release chip
    delay(3000);                               //wait for eeprom to finish writing

    Serial.println("Write Complete");


}


byte read_eeprom(int EEPROM_address)
{
    int data;

    digitalWrite(SLAVESELECT,LOW);
    spi_transfer(READ); //transmit read opcode

    spi_transfer((char)(EEPROM_address>>8));            //send MSByte
address first
    spi_transfer((char)(EEPROM_address));   //send LSByte address
    data = spi_transfer(0xFF);              //get data byte

    digitalWrite(SLAVESELECT,HIGH);         //release chip, signal
end transfer
    return data;
}

void loop()
{
      eeprom_output_data = read_eeprom(address);
      Serial.println (eeprom_output_data);
      address++;
      if (address == 128)
        address = 0;

      delay(500);                         //pause for readability
}
```

15.2.5   Experiment 4

Program to connect to external Bluetooth device

Brief explanation of setup

Bluetooth is one of the great examples for wireless connectivity. It is used in many fields. Bluetooth consumes a very small amount of energy. HC-05 communicates with the help of UART at the baud rate of 9600.

The below code illustrates to control a LED based on the command received via Bluetooth from an Android mobile. Bluetooth terminal HC-05 is one of the android mobile apps that supports to send data.

Connections:

Arduino Uno	HC-05
Rx	Tx
Tx	Rx
5v	5v
GND	GND

Firmware code

```
#include <SoftwareSerial.h>


SoftwareSerial bluetooth(2, 3);          // RX, TX


const byte CMDBUFFER_SIZE = 3;
const byte LED_PIN        = 13;          // led pin to toggle

byte ledState = LOW;
byte strLen   = 0;

void setup()
{
  // Open serial communications and wait for port to open:
  Serial.begin(9600);

  pinMode(LED_PIN, OUTPUT);


  bluetooth.begin(9600);          // Begin serial communication with
bluetooth
  bluetooth.println("Bluetooth Experiment");          // Send data
via bluetooth
}
```

```
void loop()
{
    static char cmdBuffer[CMDBUFFER_SIZE] = "";
    char c;

    if (bluetooth.available())
    {
        c = processCharInput (cmdBuffer, bluetooth.read());

        if (c == '\n')
        {
            if (strcmp("ON", cmdBuffer) == 0)
            {
                Serial.println("LED ON");
                digitalWrite(LED_PIN, HIGH);// set ledState to the pin
            }
            else if (strcmp("OFF", cmdBuffer) == 0)
            {
                Serial.println("LED OFF");
                digitalWrite(LED_PIN, LOW); // set ledState to the pin
            }
            cmdBuffer[0] = 0;
        }
    }
    delay(1);

}


char processCharInput(char* cmdBuffer, const char c)
{
    //Store the character in the input buffer
    if (c >= 32 && c <= 126)                //Ignore control characters
and special ascii characters
    {
        if (strlen(cmdBuffer) < CMDBUFFER_SIZE)
        {
        strncat(cmdBuffer, &c, 1);                      //Add it to the buffer
        }
        else
        {
                return '\n';
```

```
        }
    }
    else if ((c == 8 || c == 127) && cmdBuffer[0] != 0)          //
Backspace
    {

        cmdBuffer[strlen(cmdBuffer) - 1] = 0;
    }

    return c;
}
```

15.2.6   Experiment 5

Program to connect to external home/office Wi-Fi router
    Brief explanation of setup
    The following software code Interfaces Wi-Fi module ESP8266 to Arduino Uno which enables Internet connectivity.
    To communicate with the ESP8266 Wi-Fi module, the microcontroller needs to use a set of AT commands. The microcontroller communicates with ESP8266 Wi-Fi module using UART having default baud rate 115200 (which can be altered using AT commands).
    Connections:

Arduino Uno	Esp8266 (Esp-01)
Rx	Tx
Tx	Rx
3.3v	Vcc, CH_PD
GND	GND

Firmware code

```
#include <SoftwareSerial.h>


SoftwareSerial esp(10, 11); // RX, TX


void setup()
{
    String network = "XYZ";                      // network here
    String password = "PQR";              // password of our network
```

```
        Serial.begin(115200);
        esp.begin(115200);                    // Begin serial communication
with ESP8266
        esp.println("AT");                              // send AT command

    while (!esp.find("OK"))      // wait until we receive OK response
    {
      Serial.println("Resend");
      esp.println("AT");
    }
    Serial.println("OK Command Received");

    esp.println("AT+CWMODE=1");                    // set the ESP8266
module as a client.
    while (!esp.find("OK"))       // wait until we receive OK response
    {
      Serial.println("Setting is ....");
      esp.println("AT+CWMODE=1");
    }
    Serial.println("Set as client");

    Serial.println("Connecting to the Network ...");


    esp.println("AT+CWJAP=\"" + network + "\",\"" + password +
"\""); // connect to the network
    while (!esp.find("OK")); // wait until connected to the network

  Serial.println("connected to the network.");
    delay(1000);
}
/* Enter the AT commands via terminal and see its response */
void loop()
{
    if (esp.available())
        Serial.write(esp.read());
    if (Serial.available())
        esp.write(Serial.read());

}
```

15.2.7   *Experiment 6*

Program to connect to ThingSpeak cloud services

Brief explanation of setup

Below example explains the setup, **DHT11 sensor for sending temperature and humidity data to ThingSpeak using Arduino and ESP8266**. By this method, we can monitor our DHT11 sensor's temperature and humidity data over the Internet using the ThingSpeak IoT server. And we can view the logged data and graph over-time on the ThingSpeak website.

To view the data on ThingSpeak, firstly create a channel on ThingSpeak, and then sign up on ThingSpeak. In case if an account is already created on ThingSpeak, just sign in and enter id and password.

Connections:

Arduino Uno	Esp8266 (Esp-01)
Rx	Tx
Tx	Rx
3.3v	Vcc, CH_PD
GND	GND

Arduino Uno	DHT11
Data	GPIO pin 2
5v	Vcc
GND	GND

Firmware code

```
#include <SoftwareSerial.h>
#include "DHT.h"

#define dht11Pin      2    // Digital pin connected to the DHT sensor
#define DHTTYPE       DHT11                              // DHT 11

DHT dht(dht11Pin, DHTTYPE);
SoftwareSerial esp(10, 11);                    // RX, TX

float temperature, humidity;
void setup()
{
  String network = "XYZ";              // name of our network
  String password = "PQR";           // password of our network
```

```
dht.begin();
Serial.begin(115200);                        // Begin the terminal
Serial.println("Started");
esp.begin(115200);    // Begin serial communication with ESP8266
esp.println("AT");                           // Transmit AT command

while (!esp.find("OK"))      // Wait until we receive OK response
{
   Serial.println("Resend");
   esp.println("AT");
}
Serial.println("OK Command Received");

esp.println("AT+CWMODE=1");   // set the ESP8266 module as a client.
  while (!esp.find("OK"))        // wait until we receive OK response
  {     Serial.println("Setting is ....");
    esp.println("AT+CWMODE=1");
  }
  Serial.println("Set as client");

  Serial.println("Connecting to the Network ...");


//Command to Connect to the network.
 esp.println("AT+CWJAP=\"" + network + "\",\"" + password + "\"");
while (!esp.find("OK"));
 Serial.println("connected to the network.");

 String ip = "184.106.153.149";                //Thingspeak ip adresss
   esp.println("AT+CIPSTART=\"TCP\",\"" + ip + "\",80");   // com-
mand connect to Thingspeak.
  if (esp.find("Error"))          // check if any connection error.
  {
    Serial.println("AT+CIPSTART Error");
  }

  delay(1000);
}
```

```
void loop()
{

  delay(2000);


  for (int i = 0; i < 10; i++)
  {
    humidity = (float)dht.readHumidity();
    temperature = (float)dht.readTemperature();

    // Thingspeak command write the data. Assign appropriate write
key in the place of "xxxxxxx" to api key
    String veri = "GET https://api.thingspeak.com/update?api_key=XXXX
XXX";
    veri += "&field1=";
     veri += String(temperature);// The temperature variable we
will send
    veri += "&field2=";
    veri += String(humidity);// The moisture variable we will send
    veri += "\r\n\r\n";

    esp.print("AT+CIPSEND="); // Command to send the length of data
    esp.println(veri.length() + 2);
    delay(2000);

/*Send the data once we receive " > " */
    if (esp.find(">"))
    {
      esp.print(veri);
      Serial.println("Data sent.");
      delay(1000);
    }

  }


/* Print Humidity and temperature values on the terminal */
    Serial.print(F(" Humidity: "));
    Serial.print(humidity);
    Serial.print(F("%  Temperature: "));
    Serial.print(temperature);
    Serial.print(F("C "));
```

```
Serial.println("Close Connection");
esp.println("AT+CIPCLOSE");                        // close the link
delay(1000);
}
```

15.2.8   Experiment 7 (Fig. 15.3)

Program:

```
/*Install Mosquitto broker*/
     brew install mosquito
You will see the following if successful
     ********************************************************
  ==> Installing mosquitto
       ==> Downloading https://homebrew.bintray.com/bottles/mos-
quitto-1.4.10.el_capitan.bottle.tar.gz
     ###################################################### 100.0%
       ==> Pouring mosquitto-1.4.10.el_capitan.bottle.tar.gz
       ==> Caveats
     mosquitto has been installed with a default configuration file.
     You can make changes to the configuration by editing:
          /usr/local/etc/mosquitto/mosquitto.conf
     To have launchd start mosquitto now and restart at login:
       brew services start mosquitto
       Or, if you don't want/need a background service you can
just run:
          mosquitto -c /usr/local/etc/mosquitto/mosquitto.conf
       ==> Summary
          /usr/local/Cellar/mosquitto/1.4.10: 32 files, 618.3K
     ********************************************************
```

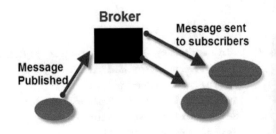

Fig. 15.3 MQTT
subscribe-publish setup

```
/*Launch Mosquitto with the configuration file mosquito.conf */
        /usr/local/sbin/mosquitto -c /usr/local/etc/mosquitto/mos-
quitto.conf

Output on the terminal will be
Mac_air-MacBook-vna:~ vna$ /usr/local/sbin/mosquito c       /usr/
local/etc/mosquitto/mosquitto.conf
        1482518967: mosquitto version 1.4.10         (build date
2021-03-31 20:09:41+12:00) starting
     1482518967: Config loaded from        /usr/local/etc/mosquitto/
mosquitto.conf.
        1482518967: Opening ipv4 listen socket on port 1883.
        1482518967: Opening ipv6 listen socket on port 1883.


/******************************************************************/
/*          On Windows
/******************************************************************/
```

Download Mosquitto for Windows and install it in the command window.
mosquito-1.4.10-install-win32.exe

Run **Mosquitto broker** by double-clicking on it. Right click and select **Start.**
Notice the status as **running**

From the command window, run:
netstat -an | findstr 1883

On the terminal, you will find:
TCP   0.0.0.0:1883   0.0.0.0:0   LISTENING
TCP   [::]:1883   [::]:0   LISTENING

Now the broker is running and listening for devices to subscribe to services. Let us assume a device drone is publishing altitude as a service.
Devices get value of altitude by using "subscribe" message as follows:
mosquitto_sub -V mqttv311 -t sensors/drone/altitude -d /* in this command, V is version, t is topic subscribed to "sensors/drone/altitude", d will enable debug

Terminal will display:
Client mosqsub/5449-vna sending CONNECT
Client mosqsub/5449-vna received CONNACK
Client mosqsub/5449-vna sending SUBSCRIBE (Mid: 1, Topic: sensors/drone01/altitude, QoS: 0)
Client mosqsub/5449-vna received SUBACK
Subscribed (mid: 1): 0

If there are no such publishers on the network, the client will start sending PINGREQ and gets PINGRESP from the broker as follows:

Client mosqsub/5449-Gastons-Ma sending PINGREQ
Client mosqsub/5449-Gastons-Ma received PINGRESP

This demonstrates that when the broker is running, any client device can publish or subscribe to services (topics) and get the required response from the broker.

The MQTT working can easily be demonstrated by graphical user interface (GUI) called mqtt.fx. Using this application, one can Subscribe, Publish and Connect, and acknowledge connection. The same is demonstrated in Fig. 15.4.

The publish command for the device is:

mosquitto_pub -V mqttv311 -t sensors/drone/altitude -m "30 f".

The following message appears on the terminal:

Client mosqsub/5532-vna received PUBLISH (d0, q0, r0, m0, 'sensors/drone01/altitude', ... (4 bytes)) 30 f

Unsubscribing is shown in Fig. 15.5.

15.2.9   Experiment 8

Instruction to access cloud services

Limited services on public cloud like Amazon Web Services (AWS) and Google Cloud (Firebase) can be accessed by registering onto their cloud sites. Limited free services on these cloud are good to familiarize with the usage of service offerings.

15.3   Part III

15.3.1   Approach Note to Develop IoMT- and IoS-Based Solution for Health Monitoring Eco Solution

IoMT and IoS ecosystem solution: Application, Architecture, Technology, and Security

As the Internet of Medical Things (IoMT) is seen as a very promising solution for the future of healthcare, adoption of the same is accelerated by the current COVID-19 pandemic. Since treatment of the COVID-19-infected person needs touch-free care, IoT system with remote care delivery is a very apt solution. Though it was not followed completely, the adoption of remote healthcare solution for diagnostics was quite common. This part explains the approach for the development of IoMT solution along with IoS and security aspects.

15.3.2   Taxonomy of IoMT Healthcare Ecosystem

Figure 15.6 shows the taxonomy of IoMT healthcare ecosystem solution.

IoMT-based healthcare ecosystem of today can be thought of as consisting of the following subsystems.

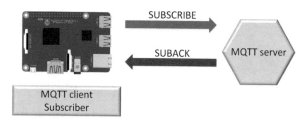

Fig. 15.4 MQTT setup

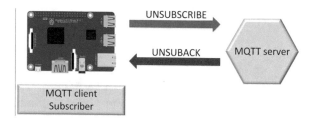

Fig. 15.5 Unsubscribing a topic

15.3.3   Health Applications

Health applications are a variety of applications which are to be developed in the form of mobile applications, video platforms, chat applications, health assist bots, and regional language conversion applications. These perform the following functionalities:

- Teleclinic
- Smart pharmacy
- Hospital management
- Security

15.3.3.1   Teleclinic

Teleclinic provides hospital services like nursing care, expert consultancy, health record interpretation, suggestion for alternative medicine, and follow-up consultancies.

15.3.3.2   Smart Pharmacy

Smart pharmacy provides services like prescription management in which logistics of delivery of prescribed medicines, dosages, and pickup services like empty cylinders, etc.

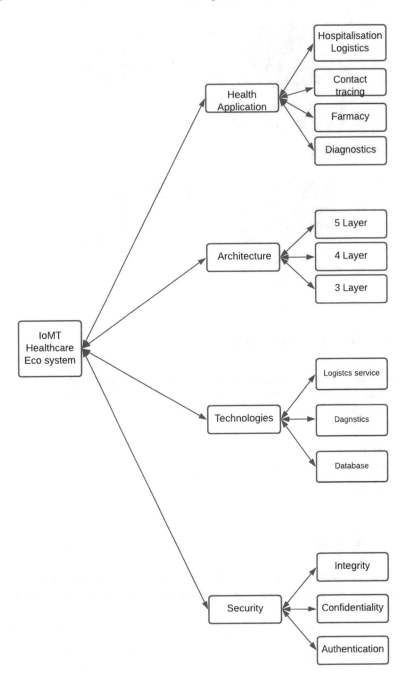

Fig. 15.6 Taxonomy of IoMT healthcare ecosystem

15.3.3.3   Hospital Management

Hospital management is logistics management services like blood collection, report collections, bed management, management of hospitalization and discharge management with follow-up visits, and insurance claims and billing.

15.3.4   Security

Security services include health record privacy, secured access to previous services, and illness and wellness records.

15.3.4.1   Architecture

Architecture of each of the applications can be integrated or stand-alone which will follow the three-layer to five-layer architecture depending on the level of proprietary ownership allowed. The solution can range from a constrained device with connectivity to only software applications. Choice of architecture has to be suitably made as per the detailed discussions in the previous book chapter.

15.3.4.2   Technologies

The IoMT-based ecosystem solution framework is shown in Fig. 15.7. *The suitable technologies for the case study are the following.*

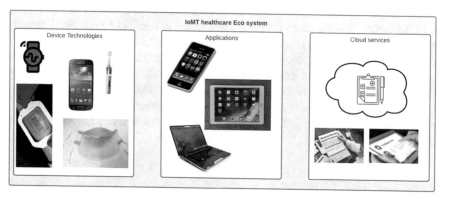

Fig. 15.7   IoMT healthcare ecosystem

15.3.4.3 Device Technologies

Device technologies are wearables, Wearable device, mobile applications, cloud service-based applications, IIoT devices, sensor tags, sensor devices, smartphone-based applications, and desktop-based applications. Devices may use technologies like RFID, sensors, actuators, biomedical patches, and special electrodes. These devices are developed using embedded design methodology discussed in earlier chapter in this book.

15.3.4.4 Smart Applications

The applications range from simple Internet-enabled mobile application to smart intelligent application based on cloud service with geo-tagging, machine learning, and AI capabilities. Smart applications for home care, vital monitoring, medicine reminders, appointment booking, subscription base doctor access, medicine logistics, home care assistance, fitness trackers, getting on-demand expert service, hospitalization logistics planning, and hospitalization finance planning are possible applications. They use smartphones and add-on devices to them to perform these functions. Most of these applications will have Internet access and can be edge devices themselves.

Some of the smart applications also target Internet of services (IoS) on the logistics and healthcare support service offerings. Access to health records, health database for policy decisions, and planning the medical infrastructures fall under IoS category.

15.3.4.5 Security

IoMT solutions are complex systems involving multiple service providers, subsystem components, and services from third parties. Hence, it is prone to security threats. Also, it deals with critical health records which are personal, and privacy is a prime factor. It is therefore necessary to comply to relevant applicable standards at different layers or subsystems and cloud servers. It is also necessary to have security features inbuilt at all levels where the system boundaries are getting crossed for communications. Security standards include authentication, privacy and data integrity standards applicable to data storage, computing at cloud and edge devices, and communication interfaces. For detailed security requirement, one can refer to the chapter presented in the IoT security which is most relevant and applicable in this case.

Other design considerations include scalability, interoperability, and regional regulatory restrictions for the IoMT-based solution proposed.

15.4 Conclusion

The realistic design case presented in this part is an approach note to trigger the design thinking while providing IoMT-based solutions for the identified problem. It is however necessary to consider the actual requirement and refer to local regulatory restrictions applicable to the solution while implementing.

Index

Printed in the United States
by Baker & Taylor Publisher Services